Essentials of Physical Activity

Second Edition

Paul Brynteson, DPE
Professor and Chair
Health, Physical Education, Exercise Science and Nutrition
Northern Arizona University

Donna Brynteson, BSN-RN
Senior Nurse
Student Health Clinic
Northern Arizona University

Fritz Huber, ED.D., C.S.C.S
Chairman
Department of Health, Physical Education and Recreation
Oral Roberts University

eb

eddie bowers publishing, inc.

14744 Highway 20 West
Dubuque, Iowa 52003 USA

**Exclusive marketing and distributor rights for
U.K., Eire, and Continental Europe held by:**

**Gazelle Book Services Limited
Falcon House
Queen Square
Lancaster
LA1 1RN
U.K.**

Book and Cover Design: David Corona Design

eddie bowers publishing, inc.
14744 highway 20 West
Dubuque, Iowa 52003 USA

ISBN 1-57879-011-5

Printed in the United States of America.

9 8 7 6 5 4 3 2 1

Contents

CHAPTER **3**

The Cardiorespiratory System   **24**

CHAPTER **4**

Cardiovascular Disease Risk Factors and Aerobic Exercise   **34**

CHAPTER **5**

The Skeletal Muscular System and Strengthening Exercises 48

CHAPTER **6**

Starting an Exercise Program for Health Fitness 58

CHAPTER **7**

Body Composition and Health Fitness   72

CHAPTER **8**

The Physically Active Lifestyle and Aging   84

CHAPTER **9**

Basic Nutrition for Health Fitness   90

CHAPTER **10**

Managing Back Injuries and Stress   **110**

Laboratory Units

122

Essentials of Physical Activity

1

Lifestyle and Health

Learning Objectives

This chapter emphasizes the importance of lifestyle to health. Historical accounts of the search for health and statistical evidence of factors that contribute to health and death are discussed. After reading this chapter you should be able to:

1. Discuss various searches for health and understand the multi-billion dollar industry surrounding that search.

2. Understand the change in leading causes of death form 1900 to 1999 and list the current most common degenerative diseases.

3. Know the four primary contributors to health and the role each play.

4. List seven lifestyle characteristics that can affect health.

Key Terms

Degenerative Diseases

Health-Care Cost

Life Expectancy

Lifestyle

Historical Search for Health

People are increasingly becoming health conscious and are seeking knowledge about sound health practices. While it may seem to be a new trend, people have always searched for health. In 1513, Ponce de Leon, at 53 years of age, organized and led an expedition in search of the fountain of youth in the New World. He believed in the legend of a fountain that, when bathed in, would restore youth, vigor, and beauty. Instead of finding it, he discovered Florida and died after he was shot by an Indian's arrow.

Centuries earlier, during the time of Confucius, Chinese emperors hired alchemists to mix doses of gold and mercury into a solution to be drunk because these metals appeared not to tarnish. Unfortunately, mercury is poisonous and caused death rather than longevity. The ancient Egyptians and Romans ate large quantities of garlic to lengthen their lives. Europeans tried a variety of roots and insects.

The ancient Greeks placed great value on health and fitness. Greek teachers emphasized exercise, good nutrition, sanitation, and discipline. The Old Testament Scriptures tied sound health practices to religious laws, especially in sanitation, cleanliness, and diet.

The late 19th century, in the United States, saw a great migration to the West. Many people went to discover gold or to start a new life, but historians estimate that at least 25 % and perhaps as high as 50 % went to the West, particularly the Southwest, in search of health.

Many people have searched for health in the form of medicines, tonics, and special foods. In the past, opium preparations have frequently been taken as painkillers with the result that large numbers of unsuspecting people became addicted. The same was true for "medicinal liquids" taken to soothe the nerves and calm the stomach. Such liquids usually contained large amounts of alcohol and often led to alcoholism in many that innocently used them. It has been reported that in the late 1800's when a current popular soft drink was first developed, it contained cocaine.

The Current Health Search

Today the search for health seems to have intensified. In "The Morality of Muscle Tone" Barbara Ehrenreich observed that the quest for health has become the new morality of the 1980's and 90's. As politicians, Wall Street, and some religious leaders have abandoned virtue, it has reappeared in "healthism," i.e. health as a transcendent value. Of the books on the weekly Top Ten best-selling list, often two or three are health-related books, usually extolling some new diet.

The estimated $50 million that was spent in the United States on diet, exercise, and health books in 1982, has grown to an estimated $120 million in 1999. However, that is a small amount compared to the total amount spent for all diet, exercise, and health-related products. The National Sporting Goods Association reported that in

1998 Americans spent $6.8 billion on sport equipment (e.g. stationary bikes, weight training equipment, golf, and etc.). Additionally, $1 billion was spent on nutritional supplements and over $5 billion on health club memberships. If you combine all expenditures on fitness, sports, and leisure areas, Americans spend more than $330 billion per year, which is more than the nation spends on national defense.

Entrepreneurs have cashed in on the population's search for an attractive, thin body that is equated to a healthy body and are promoting various products that promise to help reach that goal. Scarcely a day goes by that we don't read several of the following claims in relation to some product: "The easy way to health," "Lose weight without diet or exercise," "Recommended by a leading medical center." Although exercise has clearly been demonstrated to be beneficial to health, appearance and longevity, many individuals choose to have expensive surgical procedures performed to improve their physical appearance with little regard to their health.

Numerous diet pills, diet aids, body wraps, and electrical stimulation devices are being promoted as beneficial for rapid weight loss, one pound or more per day. If anyone is searching for health, these are definitely not ways to obtain it, since any program that recommends more than two or three pounds weight loss per week is suspect and probably doesn't work or is detrimental to health. If it does work, the weight loss is mostly water that will be rapidly regained. Advertisements that claim quick and easy ways to health should be view with skepticism.

The search for health has led to many "cures" over the centuries with some actually improving the quality of life. Unfortunately, most have had no effect on health other than to waste time and money, with some even causing death. The present consensuses of the allied health professions place the burden of health in the hands of the individual making it each person's responsibility to adopt a healthy lifestyle.

Major Health Problems Today

Health problems today are different from what they were at the turn of the century. Table 1.1 compares the leading causes of death in 1900 to those of 1997. On the one hand, the infectious diseases of 1900 have been significantly reduced or are not present on the 1997 list, and thus accounting for the increased life span. However on the other hand, the incidence of both heart disease and cancer has virtually tripled since 1900, and those diseases are currently the leading causes of death.

Today, modern medicine has eliminated most of the early deaths due to infectious diseases leading to a large increase in life expectancy from 1900 to 1950 (see Figure 1.1). With these premature deaths no longer shorting the life span of Americans, degenerative diseases have come to the forefront and have taken over as the leading causes of death. Degenerative diseases are primarily diseases of lifestyle. They often begin undetected early in life and progressively cause deterioration in health, as we grow older. Often we feel that we are in a state of health because we have no outward symptom, of disease. Sometimes the first symptom of a disease is also the last, since 40 to 50 % of all heart attack victims die before they reach the hospital after their first heart attack. Coronary arteries can be occluded (obstructed) as much as 70 % to 90 % with atherosclerosis (build up of fatty deposits on the inner walls of arteries) before any noticeable symptoms appear. This is why health is far more than just the outward appearance of freedom from disease.

TABLE 1.1

Leading Causes of Death in 1900 and 1999 in the United States

• **Causes of Death in 1900**	• **Causes of Death in 1999**
• Tuberculosis	• Heart Disease
• Pneumonia	• Cancer
• Diarrhea and enteritis	• Stroke
• Heart disease	• Chronic obstructive pulmonary disease
• Liver diseases	
• Injuries	• Accidents
• Stroke	• Pneumonia and influenza
• Cancer	• Diabetes
• Bronchitis	• Suicide
• Diphtheria	• Kidney disease
	• Cirrhosis of the liver

FIGURE 1.1

Life Expectancy at Birth from 1900 to 1997 in the United States

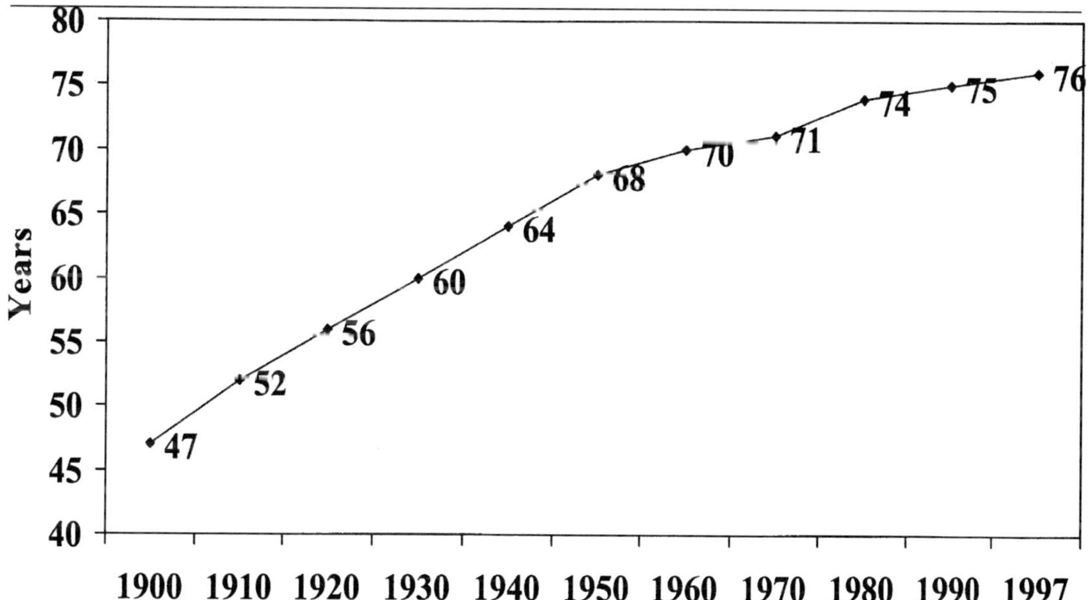

Total Health-Care Costs Increments Since 1950 in the United States

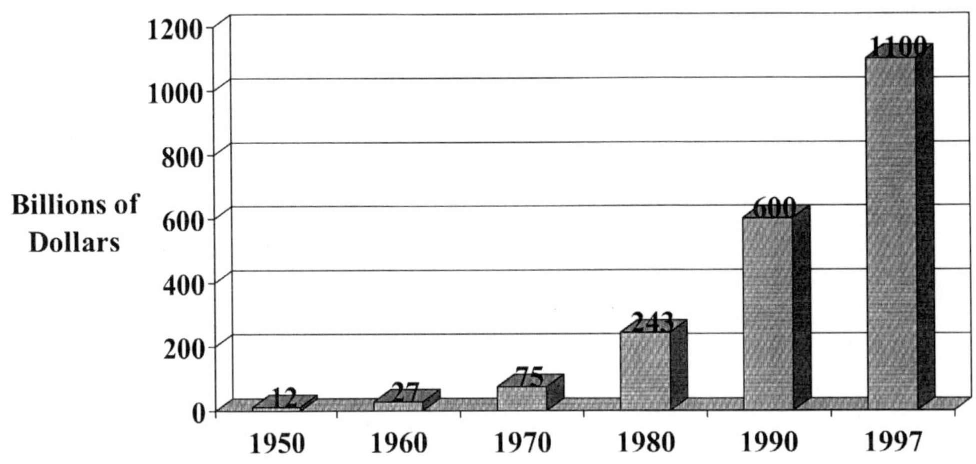

The Cost of Poor Health

In the economic arena, the effects of poor health are becoming vividly clear. In 1996, the total health-care costs topped the $1 trillion mark for the first time, up 60 % from 1990 (see Fig. 1.2). Presently, health care costs are increasing approximately $100 billion each year. The 1996 expenditure represents 14% of the federal budget and is projected to reach 18% by 2002 and 37% by 2030. By comparison, the percentage of the federal budget spent on health care is 8.6% in Canada and France, 8.2% in Germany, 6.8% in Japan, and 6.1% in England. Economists predict that by the year 2000 the health care expenditures will amount to 25% of the federal budget. General Motors spends more on health benefits for its employees than it does on buying steel for making cars. In 1990, the company added $750 to the price of each new car. Poor health is expensive for everyone. Unhealthy lifestyles are the biggest contributing factor to the staggering health-care costs in the United States.

Changing Lifestyle

The way we live today is not the way our forefathers lived. The American lifestyle has changed significantly over the last 80 years. At the turn of the twentieth century, 70 % of the American population lived in the country and was physically active in food production. They tilled the soil by walking behind a horse-drawn plow, tended animals, built their own homes, and often fought for survival. The food they ate was not processed or refined.

With advanced technology, occupations became less physically demanding. Walking behind a plow for the farmer gave way to riding on a tractor. Walking to school, church, and the grocery store gave way to riding in the automobile. Wood-burning stoves that required chopped wood for fuel evolved to automatic furnaces. Today, 95 % of our population lives in cities and is accustomed to work-saving devices, such as power lawn mowers, elevators, golf carts, and weed eaters. The new discoveries and lifestyle changes have been a mixed blessing for health.

In the years from the birth of Christ to 1900, life expectancy at birth advanced only 20 years (from 25 to 47). In the 97 years since 1900 (see Figure 1.1) however, life expectancy at birth advanced 29 years (from 47 to 76). But before drawing erroneous conclusions attributing the increased life expectancy to lifestyle, we must look more carefully at the reasons for the increased life span.

During the past one hundred years, the medical profession has made many landmark discoveries. Whereas infectious diseases such as typhoid fever, smallpox, diphtheria, tuberculosis, scarlet fever, pneumonia, measles, whooping cough, and others caused by microorganisms killed hundreds of thousands every year in the past, these diseases are almost unheard of today. Medical science has won the battle against infectious diseases through the use of vaccinations and modern medicines. In 1900 the death rate from tuberculosis was 195 per 100,000 persons, but today it is two per 100,000 persons. We used to fear outbreaks of polio but since the Salk vaccine of the 1950s, this disease no longer presents a threat.

Most of the increase in life expectancy has come from conquering infectious diseases and improvements in living conditions. Public health policies, improved sanitation, vaccinations, antiseptic surgery, and other medical discoveries have all contributed to the increase. In Figure 1.1, the slope of the curve has flattened since 1950. Ironically, the last 50 years is when most of the expensive and technologically advanced medical innovations were introduced. In reality, a person who reaches age 45 today can expect to live only two or three years longer than a person who was 45 years old in 1900.

Contributors to Health or Disease

The United States Department of Health and Human Services Centers for Disease Control and Prevention in Atlanta, Georgia has identified four factors that contribute to the cause of death and disease.

1. **Health Care.** Health care is the organization and administration of health services by professionals in our society. It involves doctors, nurses, hospitals, clinics, ambulances, and related health-care personnel and facilities. Occasionally this area is to blame for the death of a person because of the misdiagnosis of a problem, wrong prescription of medication, or some other neglect or lack of knowledge.

2. **Environment.** The evidence is clear that environment (physical, social, economic, and family) can affect our health. Living and working in high-pollution areas contribute to higher rates of emphysema and lung cancer. Other environmental problems may be stress, toxic materials, transportation, and so forth.

3. **Genetics.** A person's basic cell structure and characteristics are determined by heredity. The tendency toward heart disease, cancer, hemophilia, sickle cell anemia, diabetes and certain other diseases may be present at birth. However, heredity alone rarely causes the disease. Usually the tendency interacting with an individual's lifestyle and environment can delay or hasten the prospects of the disease.

4. **Lifestyle.** The way we live influences our health and can be a major cause of our own diseases and death. In some cases we are literally digging our own graves. Lifestyle involves total behavior, 24 hours a day, seven days a week, 52 weeks of the year. The number of sleeping hours, the food consumed, the fluids drank, the type of work, relationships with others, leisure time activities, plus all other aspects of living, reflects the individual's lifestyle.

Lifestyle and Health

The Centers for Disease Control and Prevention has evaluated the current ten leading causes of death and has allocated the proportion of each of these four contributing factors to the cause of death. Table 1.2 reveals their results.

TABLE 1.2

The Contribution of Four Factors to the Ten Leading Causes of Death in the United States

Ten Leading Causes of Death	Health Care (%)	Environment (%)	Genetics (%)	Lifestyle (%)
Heart Disease	12	9	28	54
Cancer	10	24	29	37
Stroke	7	22	21	50
Diabetes	6	0	68	26
Cirrhosis of Liver	3	9	18	70
Influenza and pneumonia	18	20	39	23
Vehicular Accidents	12	18	1	69
Other Accidents	14	31	4	51
Suicide	3	35	2	60
Homicide	0	35	2	63
AVERAGE	**8.5**	**20.3**	**21.2**	**50.3**

FIGURE 1.3

The Percentage of Contribution of Four Factors to All Deaths in the United States

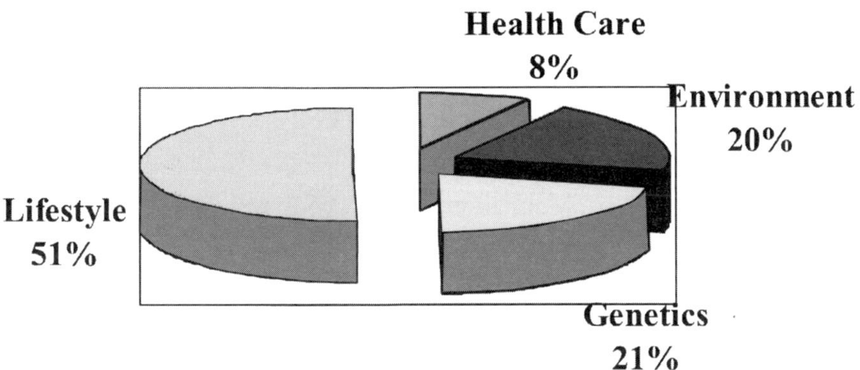

These statistics clearly place the majority of the responsibility for maintaining good health with the individual. Over 50 % of the causes of all deaths are due to lifestyle (see Figure 1.3). Heart disease and stroke are often combined into a category of cardiovascular disease. These two diseases together account for 40 % of all deaths. In cardiovascular disease, lifestyle is the major contributing factor, and when environment is combined with lifestyle, over 65 % of cardiovascular diseases have controllable components. When Joseph Califano was secretary of the Department of Health, Education, and Welfare, he said, "You, the individual, can do more for your own health and well-being than any doctor, hospital, any drug, any exotic medical device."

In addition to the statistics from the Centers for Disease Control and Prevention, other research has shown that lifestyle does significantly affect health. The American Institute for Cancer Research reports that 35 % of cancer deaths are caused by diet and 30 % by smoking. Mormons and Seventh-Day Adventists have been the target of numerous studies because of their healthful lifestyle that emphasizes good diet, exercise, avoidance of alcohol and caffeine, and positive family relationships. Mormons have a cancer rate of only 60 % of the national average, and Seventh-Day Adventists have a cancer rate of only 50 % of the national average. Additionally, Seventh-Day Adventists have one-third of the bronchitis and emphysema (both related to smoking), half of the expected heart disease, and half of the diabetes of the national average. They also have a greater life expectancy. Studies comparing death rates in Nevada and Utah revealed that the death rate in Nevada was 35 % higher for all ages than that of Utah. The primary difference between the two states was the heavy Mormon population in Utah.

Dr. Nedra Belloe and her associates have conducted an ongoing study in California of the relationship between lifestyle characteristics and life expectancy. The seven characteristics they have studied are shown in Table 1.3.

The results have demonstrated that if at age 45 you follow three or fewer of the seven lifestyle habits, your life expectancy is an additional 22 years, or age 67. If, on the other hand, you follow six or seven of the habits, your life expectancy increases by 33 years to an age of 78 years. Following these simple lifestyle habits can add 11 years to your life.

TABLE 1.3

Seven Lifestyle Characteristics that Effect Life Expectancy

- Eating three meals per day and avoiding snacks.
- Eating Breakfast every day.
- Maintaining normal body weight.
- Exercising moderately.
- Sleeping seven to eight hours per night.
- Refraining from smoking.
- Drinking alcohol in moderation or abstain completely.

TABLE 1.4

Three Cultures with Life Expectancies Over 100 Years

- Vilcabambans of Southern Ecuador
- Hunzukuts from the Himalayas of Northern Pakistan
- Caucasian Group in Russia Between the Black and Caspian Seas

Hans Kugler, in his book "Slowing Down the Aging Process", summarizes some of the key lifestyle behaviors and their health benefits for longevity. Smoking two or more packs of cigarettes per day subtracts eight to nine years from your life. You lose one year for each 10 pounds of excess weight you carry plus an additional six to 10 years off for bad nutrition. However if you exercise regularly, can add six to nine years to your life.

Today there are several known cultures that appear to have long life expectancies, many over 100 years and some surpass 120 years. All these cultures follow very similar lifestyles that include a diet low in fat, cholesterol, processed sugar, salt, and calories. In addition, the members lead physically active lives consisting of hard physical work and live in a quiet, peaceful, and relaxing environment filled with few worries, and low stress.

Life insurance companies report that 83 % of deaths before age sixty-five could have been prevented with a healthy lifestyle. Studies at the University of Tennessee and at the Massachusetts General Hospital reveal that lifestyle was a major contributing factor in more than 78 % of the hospital admissions.

Summary

The search for health dates far back into history. Currently, the health search is a multi-billion dollar industry. Effective and not so effective items for maintaining health are available to the health and fitness consumer. The cost of health care and prevalence of heart disease virtually command a change to positive lifestyle behavior. The four factors that contribute to the cause of death and disease consist of inadequate health care, the environment, genetics, and lifestyle. Lifestyle is the one factor that each individual can positively control. Life expectancy and quality of life can be improved through proper nutrition, maintaining normal body weight, exercising, adequate rest, refraining from smoking, and drinking alcohol in moderation or abstaining.

Chapter Questions

1. What personal expenditures have you made in the past year that relates to your "search for Health"?

2. What were the five leading causes of death in 1900 and 1997 and what factors have affected the change between these two periods?

3. Define degenerative diseases; list the most common ones.

4. List and explain the four primary contributors to health and disease.

5. What are seven lifestyle characteristics that can affect life expectancy?

6. What percent of cancer deaths are caused by diet and what percent by smoking?

2

Concept of Health Fitness

Learning Objectives

This chapter describes an over-all concept of health fitness and the benefits of exercise to one's health. Upon completion you should be able to:

1. Define health and understand the five health fitness components.

2. Explain the components of performance fitness.

3. Understand the concepts of the four principles of general conditioning (i.e. overload, specificity, individuality, and reversibility).

4. Explain the three general categories of exercises.

5. Understand the many health benefits of regular physical exercise to the body.

5. Discuss the relationship between lifestyle choices and health.

Key Terms

Aerobic Exercises

Anaerobic Exercises

Cardiorespiratory Fitness

Flexibility

Health

Health Fitness

Individuality Principle

Muscle Development Exercises

Muscular Endurance

Muscular Strength

Overload Principle

Performance Fitness

Reversibility Principle

Specificity Principle

What is Health

Although it is sometimes difficult to define what is meant by "good health", it is usually easy to identify the effects of "poor health". Some people experience frequent, recurring illnesses that cause them to be hospitalized. Others suffer from less severe problems that prevent them from going about their normal daily activities for a period of time. Most of us experience a feeling of being unwell at some point in our lives.

For many years health was considered by most people as "absence of disease." If a person was not sick or had no disease, the individual was said to be healthy. The current view of health is that it is more than simply the absence of disease. Suppose you view two persons, neither of whom is sick. Being free from disease, they are considered healthy. One person is overweight, is often tense and anxious, and has difficulty walking up a flight of stairs without getting out of breath. The other is of normal weight, is physically fit, has energy to enjoy numerous activities after a day's work, and is also happy and relaxed. Both have an outward appearance of freedom from disease, but they differ significantly in their quality of life and are not really in equally good health.

Just as wealth is more than absence of poverty and happiness is more than absence of sorrow, health is far more than absence of disease. Health is freedom from diseases plus the possession of physical fitness, which promotes quality living as well as longevity.

Concept of Physical Fitness

Most authorities agree that there are two general categories of physical fitness. One is fitness related to health that is called *health fitness* and the other category is fitness related to efficiency of movement and sports skill that is called *performance fitness* (see Figure 2.1).

Health fitness has five components:

1. **Cardiorespiratory Fitness.** Cardiorespiratory (CR) fitness consists of the health and fitness of the heart, lungs, blood vessels, and the blood. This system functions to deliver oxygen and nutrients to all the cells of the body and remove waste products. Efficient functioning of this system is not only important for sustaining life, but is essential for performing all physical activities. To develop the CR systems, activities must employ large-muscle groups contracting in a rhythmic manner and on a continuous basis (e.g. running, walking, swimming, and etc.).

2. **Muscular Strength**. Muscular strength refers to the ability of the skeletal muscles to exert maximal amount of force. Adequate strength is important for health in order to perform daily tasks more efficiently,

to decrease joint and muscle injuries, and to delays the weakening of muscles and bones with aging. To develop muscular strength, weight-training devices are utilized to place a workload above the amount the muscles are accustomed to.

3. **Muscular Endurance.** Muscular endurance is the ability of the skeletal muscle to perform repetitive contractions repeatedly or maintain a contraction for an extended period of time. Proper muscular endurance enhances posture, prevents low back pain, improves muscle tone, and self-esteem. To develop muscular endurance, the muscle is repeatedly contracted beyond daily levels (e.g. sit-ups, push-ups and pull-ups).

4. **Flexibility.** Flexibility is the ability to move a joint through the full range of motion without discomfort or pain. Flexibility ensures efficient body movements and is essential for preventing muscle and joint injury. To develop good flexibility, stretching exercises are performed to the point of discomfort and held.

5. **Body Composition.** Body composition refers to the amount of the body that is made up of lean tissue (i.e. bones, muscles, and body organs) and fat. Excess body fat is related to increased health problems (e.g. cardiovascular disease, hypertension, stroke, and cancer) and early death. On the other hand, too little body fat also compromises health by disrupting the reproductive system, decreasing temperature regulation, and limiting protective cushioning of internal organs and joints. To maintain healthy body composition, caloric intake and expenditure must be in balance.

The five components of health fitness are essential to good health and well being. When these components are improved, there is a decreased risk of diseases related to a sedentary lifestyle such as heart disease, Type II diabetes, and osteoporosis. One advantage of activities that promote health fitness is the fact that they may be done alone or with others. Generally, composition is only with oneself to fulfill individual goals.

Performance fitness includes the five components of health fitness, as well as the following six components:

1. **Agility.** Agility refers to the ability to move the body quickly in different directions, including stopping and starting.

2. **Power.** Power is the explosive use of strength and is similar to muscular strength except it is strength per unit of time.

3. **Reaction time.** Reaction time refers to the ability to react quickly to a stimulus, such as, moving quickly to return a fast tennis serve.

4. **Balance.** Balance is the ability to maintain equilibrium when stationary (static balance) or while moving (dynamic balance).

Components of Physical Fitness

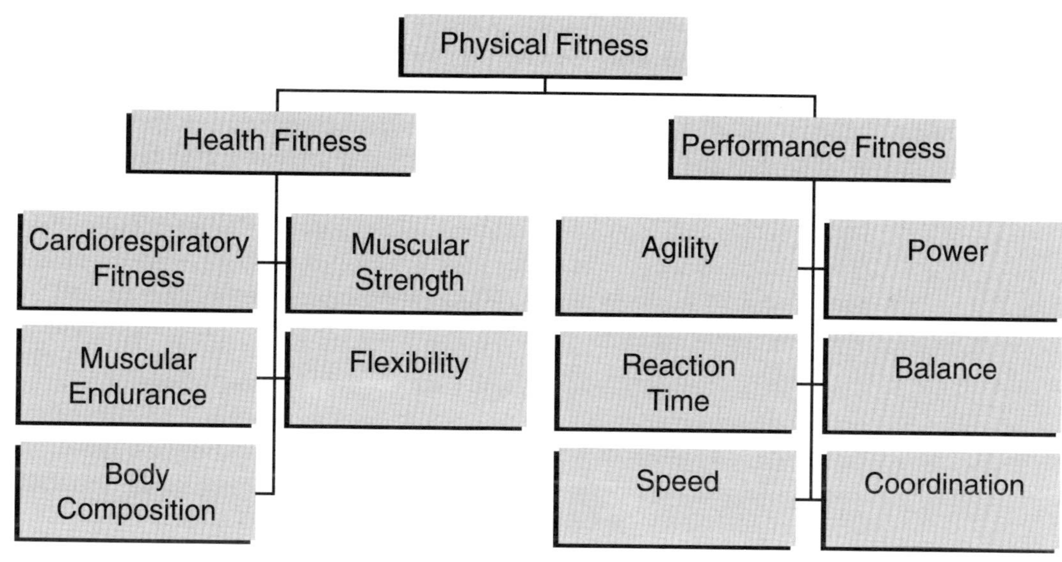

5. **Speed**. Speed refers to the quickness of movement, such as how fast one can run 50 meters.

6. **Coordination**. Coordination refers to the ability to gracefully and smoothly integrate movements of the body and different body parts such as kicking a ball, catching a ball, and performing dance steps.

These six components of performance fitness are important for an individual to successfully participate in sports activities. These components however, are not essential for a person to have good health, but the health fitness components usually do also contribute to success in sports. Many sports such as tennis, gymnastics, basketball, football, badminton, and racquetball rely heavily on both health fitness and performance fitness components. Participation in skill-related sport activities will assist in maintaining a level of health fitness, however these activities by themselves will not develop good health fitness.

Principles of Conditioning

In order to improve physical fitness a conditioning program will be necessary. The components of health fitness improve if increased demands are placed upon the bodily systems that they are dependent upon. Additionally, physical fitness is an individual matter with every person possessing different abilities and needs. Some may need to work more on flexibility while others might need to work to develop muscular strength. However, to improve physical fitness, there are four basic principles that need to be followed by everyone in order for improvement to occur.

1. **Overload**. The Overload principle states that exercising at a level of intensity higher than is normally performed will cause the body systems being stressed to adapt to that intensity and enable the body to function more efficiently. The overload principle applies equally to the development of health and performance fitness components. As important as overload is for improvement to occur, the overload must not be excessive so as to cause injury. The principle of overload is accomplished through manipulating combinations of frequency, intensity and time.

2. **Specificity**. The principle of specificity refers to adaptations in the metabolic and physiological systems depending on the type of overload being used. This implies that "you get what you train for." In other words specific exercises must be done to improve specific components of physical fitness for specific organs, muscles, and joints. For example, jogging will improve cardiorespiratory fitness but will not increase flexibility.

3. **Individuality**. The principle of individuality brings to light the many factors that contribute to individual variation in conditioning response. Genetic factors influence the conditioning response from exercise, however the general adaptation from the overload will be similar from person to person. Clearly, conditioning benefits are optimized when exercise programs are designed to meet the individual needs and capacities.

4. **Reversibility**. The reversibility principle states that the beneficial adaptations from the conditioning process are transient and reversible. Simply put, "use it or lose it". The human body is very efficient and will not maintain unused or neglected areas of the body. These areas will be eliminated or converted into another form and stored. In only a few weeks, significant conditioning benefits will be lost.

Adaptation and Progression

As a person overloads the body, it will adjust, and improvements in fitness will occur. As the body adapts to the overload, one must progressively increase the resistance or amount of work performed for continued improvement and increased health benefits. By progressing slowly, the chance of injury is limited and a person will gradually improve. Health fitness should be a lifetime goal that can not be achieved quickly and then forgotten.

Types of Exercise

Following the conditioning principle of specificity, the body will adapt according to the type of overload (i.e. exercise). There are basically four general categories of exercises, and each has specific application for benefiting different systems of the body.

1. **Aerobic**. The primary purpose of aerobic exercises is to develop the cardiorespiratory system. By placing an overload on the heart, lungs, and exercising muscles, improvements in oxygen delivery and utilization occur. These types of exercises are beneficial in preventing heart disease and improving functioning of the immune system. Aerobic exercises use large muscle groups in a repetitive, rhythmic motion for an extended period of time. They include walking, jogging, cycling, swimming, and exercise to music. Guidelines for aerobic exercises will be described in chapter 4.

2. **Muscle-development**. The primary purpose of muscle-development exercises is to develop the muscles of the body that control movement and maintain posture. These exercises include weight training using machine weights and/or free weight as well as calisthenics. They are vital in developing strength, muscle endurance, muscle tone, and good posture. Muscle-development exercises are an essential and important part of a good conditioning program. These will be discussed in chapter 5.

3. **Flexibility**. The primary purpose of flexibility exercises is to improve or maintain joint range of motion. Through appropriate stretching techniques, risk of injuries is reduced and efficiency of movement is increased. Flexibility exercises consist of stretching either alone or with a partner and require holding the stretched position for 15 to 60 seconds. Flexibility exercises should not be confused with stretching exercises used as part of a warm-up (chapter 6).

4. **Anaerobic**. The primary function of anaerobic exercises is to develop performance fitness such as speed, agility, and strength. They include sprinting, strength training, agility drills used in sports activities, and other high-intensity activities of short duration. Anaerobic exercises are especially important for sport skill performance but is not a necessary part of an adult conditioning program for health fitness (chapter 5).

Of the four types of exercise, aerobic, muscle-development, and flexibility form the core of a good health fitness conditioning program. However, all three must be included for the conditioning program to achieve health fitness.

Health Benefits from Regular Physical Exercise

The benefits of exercise upon the body depend upon the type of exercises performed, the intensity and duration of the exercise, and the number of times the exercises are done each week. Some health benefits occur immediately after performing a single exercise bout, however most require many months of consistent conditioning to reap full benefits. Listed below are some of the beneficial adaptations that can be expected when a comprehensive conditioning program is followed.

1. **Heart.** With aerobic conditioning, resting and submaximal exercise heart rate decreases. The size and weight of the heart will increase along with stroke volume (i.e. the amount of blood ejected from the heart during each contraction) during rest and exercise. This allows the oxygen needs of the body to be met with less work by the heart.

2. **Blood and its vessels**. Regular exercise will increase blood volume augmenting oxygen transport and cooling the body when in a hot environment. It can also improve the blood lipids by lowering cholesterol, low-density lipoprotein (LDL – bad cholesterol), and triglycerides. In addition, high-density lipoprotein (HDL - good cholesterol) will increase. The blood vessels as well improve by having greater elasticity and distensibility (ability to expand or stretch). Because of this, resting and submaximal blood pressure decreases and lessens the workload of the heart. In addition, the blood vessels increase in cross-sectional area and create more capillaries which function to deliver more oxygen to cells, thus enabling all the body's cells to function better. Also, the blood vessels will increase their ability to dilate and constrict allowing delivery of the blood to areas where the need is greatest at any given time.

3. **Pulmonary system**. Regular exercise improves the functioning of the pulmonary system in several ways. The lungs will have a greater capacity to inhale and exhale and will also allow more oxygen to be diffused into the blood. During submaximal exercise, conditioned individuals breath less than before they began a conditioning program. This improved ventilatory economy increases the oxygen availability.

4. **Skeletal muscles**. Aerobically conditioned muscles will utilize more fat for energy than unconditioned. Strength-trained muscles will be stronger and more resistant to fatigue. Also hypertrophy (increase in cross-sectional area) will occur, however the amount is dependent upon type of training, gender, genetics and maturation. Tendons and ligaments also increase in strength, decreasing joint injuries and improving the function of an injured joint. The development of muscle tissue is a significant factor in slowing down the aging process.

5. **Body Composition**. Both aerobic and strengthening exercises work to decrease excess body fat. Through aerobic conditioning the body improves its ability to mobilize fat from fat stores and use it for energy and increases the amount of fat used in a 24-hour period. Weight training on the other hand, increases muscle mass, which is more metabolically active than fat tissue, causing increased caloric expenditure over the day.

6. **Bones**. Regular exercises consisting of weight baring and strengthening activities increase calcium deposition in the bones being overloaded. Weak bones lead to osteoporosis (porous and brittle bones) which affects 25 million women in the United States and is responsible for 250,000 hip fractures each year at a cost of $10 billion. Calcium supplementation alone will not decrease this disorder; exercise is needed.

7. **Psychological**. A regular exercise program favorably modifies the psychological state. These benefits include a reduction in anxiety, a decreased level of depression, and an improved self-esteem. Exercise negates the harmful effects of stress. In addition, regular conditioning can greatly improve ones personal appearance, enhancing self-concept and body image.

8. **Immune system**. Aerobic conditioning improves the natural immune function of the body. Natural Killer cells increase their activity in inactivating viruses and the metastatic potential of tumor cells. In addition, aerobic exercising slows the age-related decrease in T-cell function, which defends against viral and fungal infections, and assists in regulating other immune mechanisms. Also, regular exercise develops the body's antioxidant functions and increases intestinal transit time. However, high intensive exercise or a rapid increase in training has the opposite effect and decreases the body's immune system. Numerous studies have shown that physically fit persons are sick less often, see their doctors less, are hospitalized less, and spend less on medical care than persons who are not physically fit.

The *Journal of the American Medical Association* published a major study in 1989 where 13,344 people were followed over an average of eight years. Figure 2.2 shows the results indicating that the fit males and fit females had a significantly lower death rate than the unfit males and unfit females from cancer, cardiovascular disease, and all diseases combined.

A study from the Massachusetts General Hospital reported that 86 % of hospitalizations probably could have been prevented if patients had followed a healthful lifestyle. Separate studies done on employees of Northern Natural Gas, Tenneco, Mesa Petroleum Company, Allen Bradley Co., New York State Department of Education and Civil Services, NASA, and Purdue University reported less sickness,

Death Rates for Fit and Unfit Males and Females

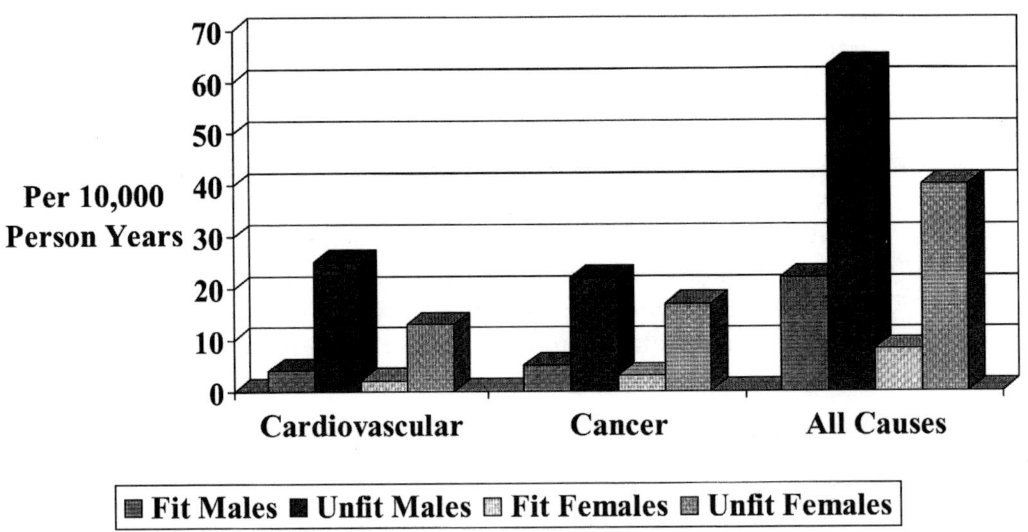

better health, better job performance, and fewer health insurance claims among employees who were physically fit. Businesses are beginning to realize that they can't afford to hire unfit employees. In the last ten years more than three hundred companies have started fitness programs for their employees.

Exercise and Aging

Regular physical conditioning delays or slows the aging process (see Figure 2.3). For years, the decrease in physiological functioning of the aging body was considered to be the normal result of growing old. However, recently researchers have started to separate poor health fitness from the normal physiological decline due to aging. Even if a sedentary lifestyle has been followed for decades, starting a conditioning program will generate benefits and slow the aging process. Studies have shown that even into the ninth decade of life, the body will adapt and improve with a health fitness conditioning program. The sooner an inactive individual begins to exercise (after medical clearance from a physician), the greater the delay in the aging process and the sooner an increase in health benefits. However, it is never too late to begin an exercise program. Even after decades of sedentary living, possibly including a heart attack, a healthier and more productive life can be gained.

Geneticist Leonard Hayflick has stated that on the basis of his research on cell division and deterioration, mankind could expect a life span of 110 to 120 years. He

Physiological Function of Active and Sedentary Individuals as Compared to a Sedentary 20 Year Old

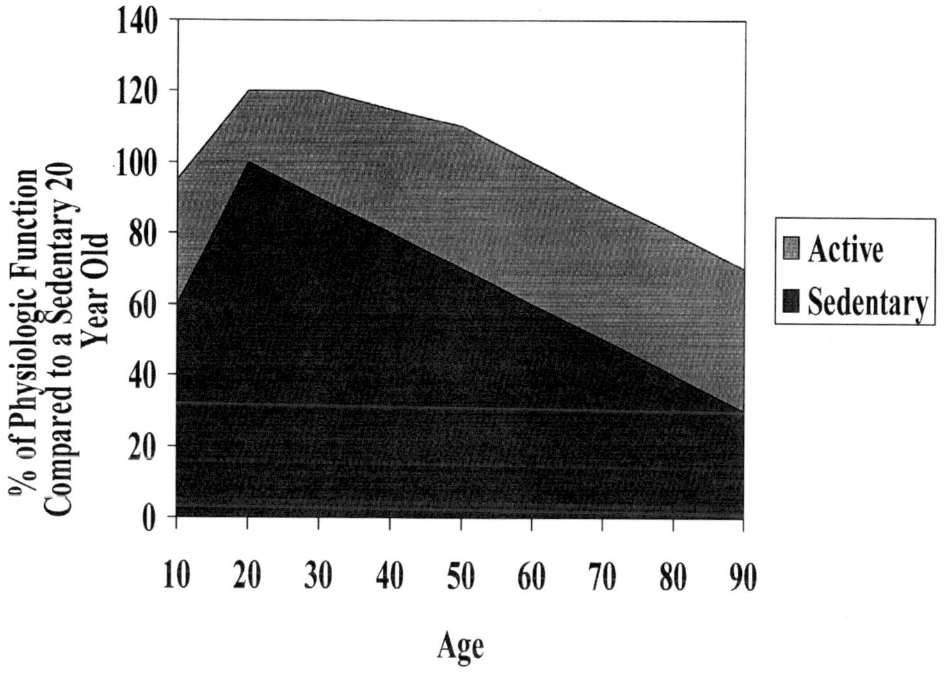

reports that human cells in a perfect environment continue to divide until they stop dividing and die according to their genetic code. Evidence is mounting that the way people live their lives is significantly affecting health and life expectancies.

Summary

Health includes absence of disease and the possession of sufficient physical fitness that allows one to live a full life of quality and quantity. Health fitness includes cardiorespiratory fitness, muscular strength, muscular endurance, flexibility, and good body composition. Performance fitness includes agility, power, reaction time, balance, speed, and coordination. To improve health or performance fitness, the principles of overload, specificity, individuality, and reversibility must be followed. Exercises for improving fitness include aerobic, anaerobic, muscle-development, and flexibility activities. A regular conditioning program will benefit all systems of the body and as a result improve the quality of life.

Questions

1. Discuss the statement "Health is the absence of disease."
2. List and define the five health fitness components.
3. List and define the six performance fitness components.
4. Explain the four principles of conditioning.
5. List and explain the four general categories of exercise, and discuss the benefits of each to different parts of the body.
6. Discuss at least six health benefits from regular physical exercise.
7. Which type of blood lipid is increased with aerobic exercise?

3

The Cardiorespiratory System

Learning Objectives

This chapter explains the anatomy and physiology of the cardiorespiratory system and its relationship to health fitness. Upon the completion of this chapter you should be able to:

1. Know the four basic needs of cells.

2. Understand the four components of the cardiorespiratory system.

3. Understand the significance of systolic and diastolic blood pressure.

4. Describe the function of the lungs in exercise.

5. Understand the importance of hemoglobin.

6. Describe the four methods that help blood return to the heart.

Key Terms

Arteries

Arterioles

Atrium

Blood Pressure

Capillaries

Cardiac Output

Coronary Vessels

Diastolic Pressure

Heart Rate

Hemoglobin

Hypertension

Ischemic

Myocardium

Stroke Volume

Systolic Pressure

Varicose veins

Veins

Ventricle

Requirements of the Cell

The basic unit of life in the human organism as well as in all other life forms is the cell. More than one trillion cells are in the human body. It would take about 40,000 blood cells to fill the typewriter letter O. Every minute about three billion cells in our body die, and three billion new cells are formed to replace them.

Every cell has a specific function. Some cells are specialized to cause movement (muscle cells), some are specialized to carry messages (nerve cells), and others are specialized to carry on other functions such as digestion and reproduction. However, for all cells to function, regardless of their structure and purpose, there are four requirements: 1) every cell needs oxygen; 2) every cell needs nutrients (carbohydrates, proteins, fats, vitamins, and minerals); 3) every cell needs water; and 4) every cell needs to get rid of waste products. Of the four requirements, the need for oxygen is the most critical since the body can survive for weeks without food and days without water but only minutes without oxygen.

The Cardiorespiratory System

The cardiorespiratory system functions to deliver a constant supply of oxygen to all the cells of the body and to remove carbon dioxide. The four components of the cardiorespiratory system are 1) heart, 2) lungs, 3) blood, and 4) blood vessels.

1. Heart

The heart is a muscular pumping organ with four chambers (see Figure 3.1): two upper chambers (right and left atriums) and two lower chambers (right and left ventricles). The heart functions as a double pump that provides the force to transport blood through the body to every cell. Each time the heart muscle (myocardium) contracts, it creates a pressure that forces blood into the arteries to deliver oxygen and nutrients to the cells. The heart has a tremendous capacity to vary in the number of times it contracts (i.e. beats) per minute. An adult's heart rate may vary from 40 to 80 beats per minute at rest to as high as 160 to 200 beats per minute during maximal exercise.

The resting and submaximal heart rate varies considerably between fit and unfit persons. This is due to the strength of contraction of the myocardium. Table 3.1 shows that a highly fit person's heart will beat almost half as much as an unfit person's heart. The fit (conditioned) heart may beat as much as 43,200 fewer times per day than that of the unfit heart. Obviously the slower the heart beats, the better the myocardium itself can receive blood to feed its cells. The myocardium receives its own blood supply only when it is at rest and not contracting.

FIGURE 3.1

Anatomy of the Human Heart and the Direction of Blood Flow

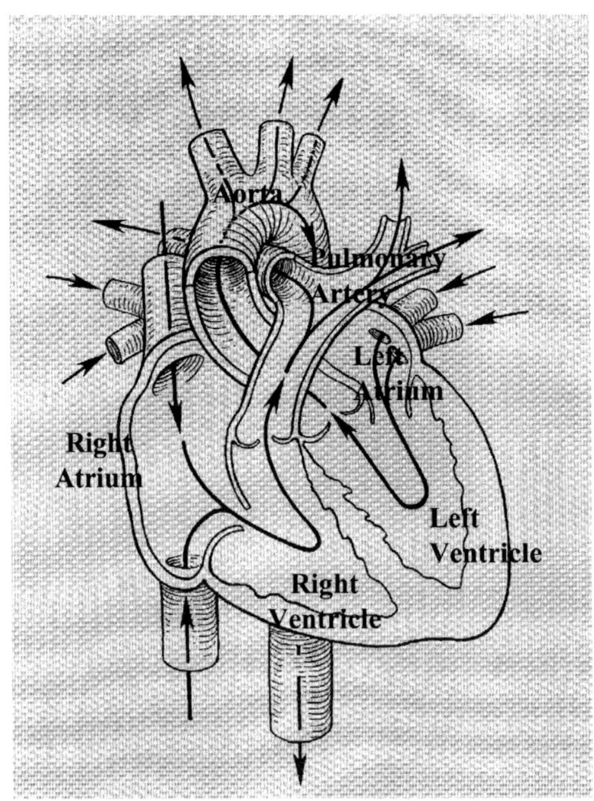

Reprint with permission from Williams & Wilkins

TABLE 3.1

Resting Heart Rate Per Minute Health Fitness Standards for an Adult Without Heart Disease

Health Fitness Standard	Heart Rate Per Minute
Excellent	< 60
Good	60 - 65
Average	66 - 70
Poor	71 - 80
Very Poor	> 80

TABLE 3.2

Cardiac Output at Rest and Maximal Exercise for a Fit and Unfit 20 Year Old

	Heart Rate	X	Stroke Volume	=	Cardiac Output	Activity Level
Unfit	75 bpm		67 mL		5 L	
Fit	50 bpm		100 mL		5 L	**REST**
Unfit	200 bpm		100 mL		20 L	
Fit	199 bpm		160 mL		32 L	**MAXIMAL**

The stronger the heart muscle, the more blood that can be pumped and delivered to the cells of the body. The amount of blood pumped from the heart per contraction is called stroke volume. When the stroke volume is multiplied by the heart rate per minute, cardiac output is obtained. Cardiac output reflects the functional capacity of the circulation to meet the demands placed upon it. At maximum workloads, a highly fit heart can pump 25 to 30 quarts of blood per minute, whereas an unfit heart may pump eight to ten. However, cardiac output at rest and during submaximal exercise is similar in both the fit and unfit. As shown in Table 3.2, the fit heart has a greater stroke volume allowing a slower heart rate as an indicator of cardiorespiratory fitness.

Coronary Blood Flow. Although tremendous quantities of blood are pumped through the atria and ventricles every minute, the heart itself does not receive any nutrients or oxygen from this blood flowing through its chambers. Rather, the heart is dependent on its own vascular system call the coronary vessels.

The coronary arteries branch off the aorta and supply the myocardium with blood. The myocardium requires a constant supply of oxygen, and when the body is at rest it utilizes up to 80 % of the oxygen carried in the blood flowing through the coronary vessels. During increased rate of contraction (e.g. exercise or stress), blood flow through these arteries must also increase to meet the oxygen needs of the heart. When fat filled plaque deposits occur in the coronary arteries (i.e. atherosclerosis), the flow of blood to the myocardium is reduced. This presents no problem at rest even with up to 80 % blockage, however as greater demands are placed upon the heart, the oxygen delivered becomes insufficient.

When the demand of the myocardium for oxygen is not met by the coronary blood flow, the area of the heart deficient in oxygen becomes ischemic (i.e. lack of oxygen due to inadequate blood supply). This lack of oxygen can result in angina pectoris, a pain in the chest that usually radiates to the left shoulder and arm, a warning signal of impending heart damage. Ischemia to the myocardium can lead to death of the affected area, called a myocardial infarction, which is more commonly know as a heart attack. The severity of the heart attack depends upon the location and size of the infarction. The heart attack may be unnoticed by the individual, or it may result in their death.

2. Lungs and Breathing

The major function of the lungs is to provide a location for gas exchange between the blood and air. Within the lungs are over 300 million microscopic sacs called alveoli where air is located inside and blood is on the outside. This presents a great surface area for gas exchange; if the alveoli where opened and laid flat they would cover half a tennis court. During respiration, oxygen moves from the inhaled air through the alveoli and into the blood. At the same time, carbon dioxide from the blood diffuses into the alveoli and out of the lungs on exhalation.

As the inspired air moves through the passageways to the lungs it is moistened, filtered of foreign particles, and brought to body temperature. Breathing is an involuntary response regulated by several systems, however the higher brain centers can override allowing for temporary voluntary control. Normal healthy lungs are not a limiting factor in exercise. These organs are very efficient at moving large volumes of air into and out of the body, as well as conducting gas exchange with the blood. Smoking (e.g. tobacco and marijuana) or living in a high pollution environment including one with second-hand smoke will cause damage and ruptures to the alveoli decreasing the efficiency of the lungs and possibly causing early death.

The diaphragm and intercostals are the chief muscles involved in breathing. During exercise, these muscles must work more intensively, and for unfit individuals the constant contraction and relaxation of the diaphragm may lead to a sharp pain in the side of the lower chest. This side ache is caused by a lack of oxygen to the diaphragm muscle and will slow persons exercise intensity. However, this muscle will adapt as the body becomes conditioned, and soon this side pain will no longer occur.

3. Blood Vessels

The blood vessels of the body form the circulatory system through which oxygen and nutrients are transported to all the cells of the body and carbon dioxide is removed. This transporting system consists of three different categories of blood vessels: 1) arteries, 2) capillaries, and 3) veins.

Arteries are tubes that carry blood away from the heart and are divided into groups called large arteries, small arteries, and arterioles. Large arteries are able to withstand high pressures due to the walls being composed of several layers of connective tissue and smooth muscle. Even though these walls are impervious to gases because of their thickness, they are still elastic but only stretch slightly under high pressure. The large arteries branch into smaller arteries.

The small artery walls have similar construction and characteristics as the large arteries but have fewer layers of connective tissue and smooth muscle. These walls are still too thick for gases to penetrate. The small arteries branch further into smaller vessels called arterioles.

The walls of the arterioles have smooth muscle circling the vessels and function to regulate regional blood flow. This is accomplished by contracting or relaxing the smooth muscles circumferencing the vessel causing the internal diameter to be altered. During exercise the arterioles leading to the liver and digestive system constrict causing a reduction of blood flow to these areas, while at the same time, the arterioles leading to the skin and working skeletal muscles are dilated causing an increase flow. By redistributing blood flow to those areas in greatest need (i.e. exercising muscles), an increase in oxygen is made available. Also with additional blood flow to the skin, the extra heat generated by the contracting skeletal muscles will be better able to dissipate.

The second category of blood vessels is called capillaries. Capillaries are microscopic, thin walled tubes that connect arteries to the veins. This thin wall consists of a single layer of endothelial cells, which allow oxygen in the blood to diffuse into the cell and carbon dioxide from the cell to enter the blood. The capillaries are very narrow vessels, causing the red blood cells to squeeze through in single file. When combined with the low pressure in the capillaries, a red blood cell's transition time through these vessels takes approximately 1.5 seconds. This produces an extremely effective method for gas exchange between the blood and tissue. The density of capillaries in human skeletal muscle is over 2,000 per square millimeter of tissue. Similar capillaries are located in the lungs and lie side by side with the alveoli for gas exchange in the lungs.

The final category of blood vessels is the thinned walled veins, which guide the blood back to the heart. Small veins called venules collect the blood from the capillaries and empty into progressively larger vessels (veins) until reaching the heart. The veins are constructed of less smooth muscle and connective tissue than the arteries making them very expandable but not very elastic. The low pressure of the venous blood requires additional assistance to return the blood back to the heart. This is especially needed when the body is in an upright position and gravity is pulling the blood down into the legs.

Returning Blood Flow.   There are four methods the body uses to help return blood in the veins back to the heart. The first is a series of flap-like valves spaced at short intervals in the veins. These thin, membranous valves permit a one way blood flow toward the heart. Sometimes however, valves in a vein become defective and no longer maintain the one-way blood flow. This condition is called varicose veins and generally occurs in the surface veins of the legs. Persons who stand motionless or are on their feet for long periods of time, such as dentists, nurses, and checkout counter persons are susceptible to pooling of blood in the veins, leading to varicose veins. This disorder can become painful and lead to other problems requiring surgery.

The second method for assisting venous blood to the heart is the active contraction and relaxing of skeletal muscle. Because the walls of the veins are thin, skeletal muscle contracting in a rhythmic action squeezes the veins to propel the blood similar to milking a cow. After exercising (e.g. running), an active cool down (e.g. walking) should be performed during the recovery to facilitate blood flow to the heart. By standing still after exercising, blood will pool in the legs due to gravity and dilated arterioles, decreasing blood flow to the heart and brain sometimes causing the individual to become faint.

Smooth muscle wrapping the veins also assists blood flow back to the heart and is the third method. This smooth muscle contracts and relaxes causing a "milking" action similar to the skeletal muscle in pushing the blood to the heart against gravity. However, this method is not as effective as compared to the squeezing action of the skeletal muscle due to the small amount of smooth muscle involved.

The final method the body uses to help return blood to the heart is the change in thoracic pressure. During respiration, intrathoracic pressure decreases with inspiration and increases with expiration. This change in pressure compresses and expands the thin walled veins located in the thoracic cavity. Once again, a "milking" action is created and forces blood to the heart. During weight lifting and other straining type activities, intrathoracic pressure can be increased greatly by holding ones breath, or more accurately called the Valsalva maneuver.

Valsalva Maneuver.   The Valsalva maneuver is a forced exhalation against a closed glottis and is commonly performed while lifting heavy objects or straining. This procedure stabilizes the trunk and increases the action of muscles attached to the chest. Unfortunately, this increase in intrathoracic pressure collapses the thin walls of the veins located in the thoracic region and greatly reduces blood returning to the heart. Several physiological consequences occur due to performing this maneuver. First blood pressure rises abruptly and increases the strain on the heart. For an individual with an underlying heart disease this could be fatal. Following the abrupt rise in blood pressure is a sharp decrease in blood flow to the heart and a drop in blood pressure below resting values. Due to this low pressure, blood flow to the brain is diminished causing dizziness, seeing spots, and fainting. All of these negative consequences can be lessened by not holding the breath but exhaling during the straining activity.

Blood Pressure.   The ejected blood from the left ventricle has a tremendous force behind it that creates pressure against the artery walls and is called systolic blood pressure. This surge in pressure causes the arterial wall to distend and snap back to its original size, causing the characteristic pulse felt in the superficial arteries (e.g. carotid and radial). The systolic blood pressure represents the work of the heart.

TABLE 3.3

Health Fitness Standards for Resting Systolic and Diastolic Blood Pressure

Health Fitness Standard	Systolic	Diastolic
Excellent	105 -115	60 - 72
Good	116 - 125	73 - 80
Average	126 - 135	81 - 88
Poor	136 -150	89 - 97
Very Poor	> 150	> 97

When the heart relaxes, the pressure against the artery walls decreases as the blood flows away from the heart into the arterioles and indicates peripheral resistance. This pressure is called diastolic blood pressure and measures how quickly the blood flows from the arterioles into the capillaries.

Average values for blood pressure range widely (see Table 3.3). Blood pressure usually increases with age, and this increase is often accepted as a normal part of aging. However, this increase is not desirable because it may indicate that the vascular system is providing greater resistance to the blood flow due to inelasticity of the arteries.

Normal resting blood pressure is considered 120/80 (systolic pressure/diastolic pressure), but 100/70 and 130/85 are also normal. Blood pressure varies throughout the day according to activity, stress, and anxiety levels. It is also directly related to cardiac output and arterial resistance to blood flow. High blood pressure (i.e. hypertension) is called the silent killer because there are no outward signs or symptoms and can go undetected until a stroke, heart attack, or kidney failure occurs. Therefore, it is extremely important that blood pressure is taken regularly and if it is high, control it through diet, sodium restriction, aerobic exercise, relaxation techniques, and if necessary, medication.

If we consider only resting blood pressure, a 35-year-old person with normal resting blood pressure of 120/80 can expect to live to age 76 years. On the other hand, a person with resting blood pressure of 150/100 or greater can expect to live only to 60 years of age. This represents 16 years of reduced life expectancy. Individuals should know their blood pressure and, if necessary, take the corrective lifestyle changes to lower it.

4. Blood

The oxygen is carried in the blood by the red blood cells and is specifically attached to the iron portion of the hemoglobin of the red blood cell. Therefore, it is critically important to have a good source of iron in your diet. The normal hemoglobin for an adult male is about 15 grams per 100 milliliters of blood at sea-level conditions and for females it is about 14 grams per 100 milliliters. Low iron intake causes a reduced concentration of hemoglobin leading to a disorder called iron-deficiency anemia. Symptoms range from sluggishness to a reduced capacity to perform even low levels of exercise. Women are at a higher risk than men for developing iron-deficiency anemia due to iron loss during menstruation. Women on a vegetarian diet are at an even higher risk since the iron contained in vegetables are not as readily absorbed by the body as compared to the iron from meat sources. Inadequate daily iron intake occurs frequently among 30 to 50 % of females in the United States.

Water is the single largest constituent of the blood and makes up approximately 50 % of its volume (three to four liters). Water in the blood is needed to maintain cardiac output and blood pressure, as well as a medium to transport nutrients and remove waste. In addition, water from the blood is used as the major source of sweat to cool the body. Each day the body losses water through perspiration, respiration, urine, and feces. If this water loss is not replenished, a dehydrated state occurs leading to strain on the circulatory system and poor thermoregulation.

One of the benefits of an exercise program is that it will increase one's plasma volume. This elevated blood volume enhances both the transporting of oxygen and the regulating of body temperature.

Summary

The cell is the basic unit of life in the human organism. Cells have varied functions, but all cells require oxygen, nutrients, and water, and must get rid of waste products. It is the responsibility of cardiorespiratory system to fulfill these needs of every cell. The cardiorespiratory system is made up of the heart, lungs, blood vessels, and blood.

Questions

1. What are the four survival requirements of cells?

2. List the four components of the cardiorespiratory system.

3. Define cardiac output and distinguish between a fit person's heart and an unfit person's heart.

4. Describe the differences between systolic and diastolic blood pressure.

5. What blood pressure reading is considered average?

6. Why is hypertension referred to as the "silent killer?"

7. Distinguish between arteries, veins, and capillaries.

8. Diagram and label the structures of the heart.

4

Cardiovascular Disease Risk Factors and Aerobic Exercise

Learning Objectives

This chapter examines the controllable and uncontrollable risk factors for developing cardiovascular disease. In addition, information will be given on aerobic type exercises that are needed to improve the cardiorespiratory system and decrease the incidence of cardiovascular disease. When you complete this chapter you should be able to:

1. Explain the process of atherosclerosis.
2. Describe the controllable and uncontrollable risk factors of cardiovascular diseases.
3. Know the consequences of high blood pressure and how it can be controlled.
4. Understand what makes an exercise aerobic.
5. Determine your own heart rate training zone for aerobic exercise.
6. Understand the components of the overload principal for developing cardiorespiratory fitness.
7. Analyze various aerobic activities in terms of their strengths and weaknesses.
8. Describe the physiological and psychological adaptations from aerobic conditioning.

Key Terms

Aerobic Exercise

Atherosclerosis

Cardiovascular Disease

Cholesterol

Controllable Risk Factors

Diabetes

Frequency, Intensity, and Time

Heart Rate

High-density Lipoprotein

Hypertension

Low-density Lipoprotein

Uncontrollable Risk Factors

Cardiovascular Disease

The term cardiovascular disease includes more than 20 different diseases that affect the heart (cardio) and arterial blood vessels (vascular). Heart disease and related cardiovascular problems cause 50 % of all deaths in the United States. For the year 1997, the American Heart Association estimated the total cost of cardiovascular disease at $259 billion (see Figure 4.1). The underlying cause in 95 % of cardiovascular diseases (e.g. heart attack, stroke, and hypertension) is atherosclerosis. The atherosclerotic process occurs in the following manner:

1. Chemical changes of various compounds and oxidation of cholesterol in low-density lipoprotein occur.

2. Lesions (fatty streaks) begin to protrude on the inside of the arterial wall composed of several lipids (cholesterol, triglycerides, and phospholipids).

3. Minerals, especially calcium, also deposit on the inner layer of the arterial wall and are called plaque.

4. The artery narrows, decreasing blood flow through the vessel and to the heart or brain.

FIGURE 4.1

Costs Related to Cardiovascular Disease in 1997

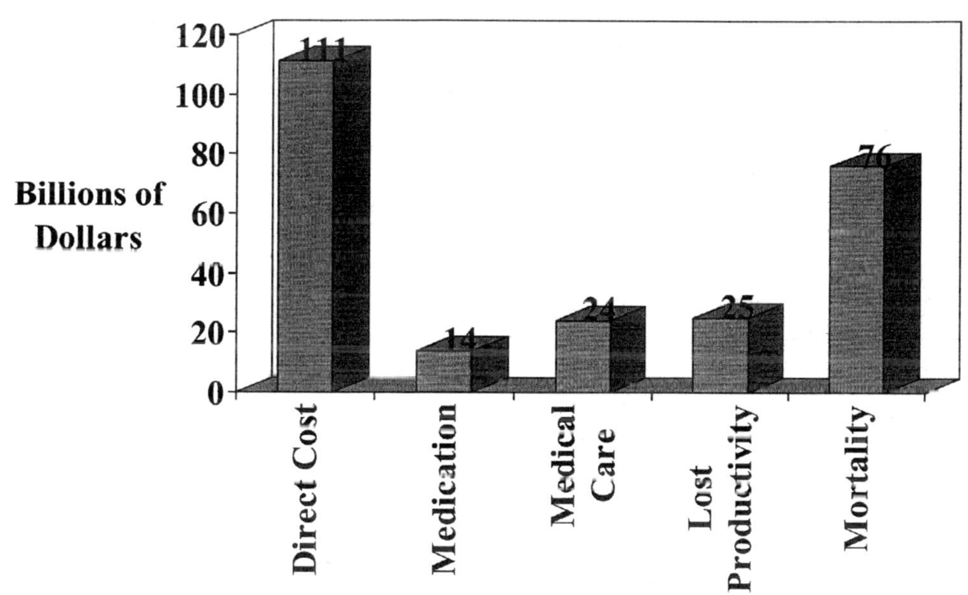

Steps Leading to a Heart Attack or Stroke

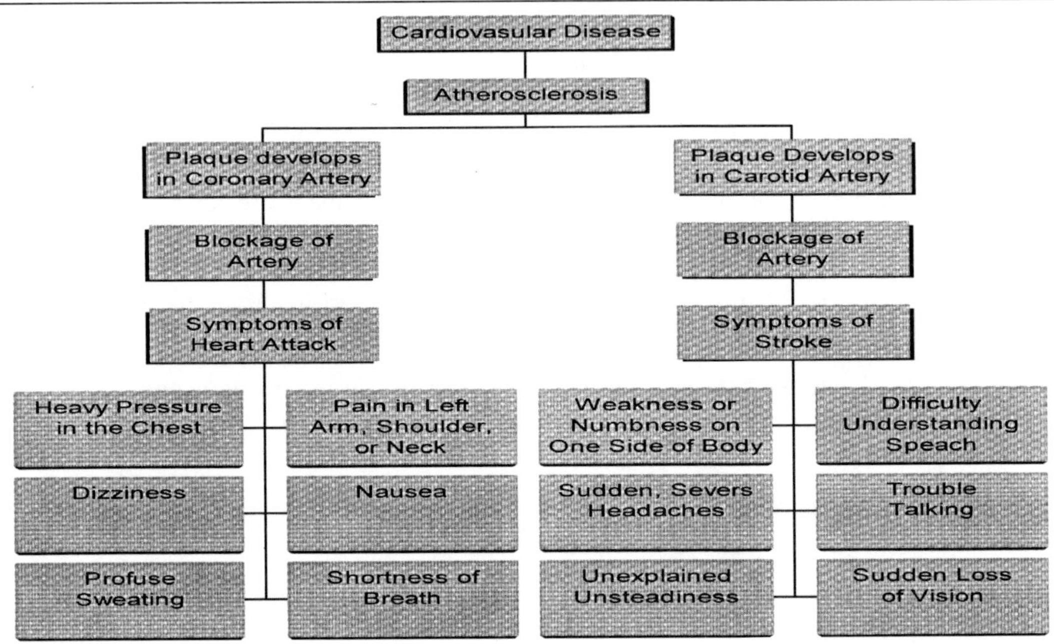

The atherosclerotic process goes unnoticed without symptoms during rest and low levels of activity until the arteries are 80 % occluded (closed). At this point, if the occluded artery is in the heart, angina pectoris may be experienced during physical activity or periods of stress. If the occluded artery is in the brain, brief blackouts, dizzy spells, or numbness of the face may be experienced (see Figure 4.2).

At one time authorities considered atherosclerosis to be a part of aging and an old person's disease. Medical evidence discovered over the past 30 years reveals that atherosclerosis is not just an old person's disease but begins in the very young and progresses. Autopsies done on Americans killed in the Korean War revealed that 77 % had some degree of coronary atherosclerosis. Unfortunately, recent studies have found that by age five, fatty streaks are common in the coronary arteries of American children.

As the arteries continue to narrow because of atherosclerosis, a time will come when blood cannot flow through the artery resulting in a heart attack or a cerebral vascular accident. If the blocked artery supplied a very small part of the heart or brain, the heart attack or stroke may go unnoticed. That results in a "silent heart attack" or a "transient ischemic attack" in the brain. If the artery supplied a larger portion of the heart or brain, the person may experience any or all of the symptoms as described in Figure 4.2. If the artery is supplying a major portion of the heart, death may occur immediately as it does in 40 to 50 % of the cases.

Risk Factors for Developing Cardiovascular Disease

Many factors have been identified that contribute to the quickening of the atherosclerosis process and cardiovascular disorders. They basically fall into two categories: 1) those that cannot be controlled and 2) those we can control either by ourselves or with the assistance of a physician.

1. Uncontrollable Factors

Heredity.   A history of cardiovascular diseases in a family means increased risk for the disease. A history usually means family members (parents, grandparents, brothers, and sisters) have died from some cardiovascular disease before age 60. Influencing factors may be an inherited genetic predisposition of the body to deal effectively with blood fats. Some families have such a strong inherited genetic weakness that few males have lived past age 40.

Gender.   Statistics reveal that cardiovascular disease is more prevalent in men than in women, especially before age 45. This is probably due to the protective function of the female sex hormone, estrogen, in delaying the atherosclerosis process. The amount of estrogen after menopause is significantly reduced; thus its benefit is not as great for females after age 45.

Age.   Growing older increases the chances of developing cardiovascular diseases. Sixty percent of deaths to persons over the age 65 years are due to heart attacks as compared to only 11 % of deaths in persons 15 to 24 years of age.

2. Controllable Factors

Fortunately, research has identified factors related to cardiovascular diseases that can be controlled. These factors are related to lifestyle and make the greatest contribution toward reducing these diseases. Key factors are briefly summarized here and are discussed in greater detail throughout the book.

High Blood Pressure (Hypertension).   More than 30 % of all Americans have high blood pressure (chapter 3) with approximately 38 % of the blacks and 29 % of the whites affected. Hypertension may begin in childhood or during the adult years. It is a silent disease, often with no symptoms and if left untreated, it greatly increases the risk of heart disease, stroke, and kidney failure. The Centers for Disease Control and Prevention report that while 66 % of hypertensives where aware of their high blood pressure, only 24 % had it under control. The earlier in life the problem starts, the more severe the consequences. High blood pressure can be controlled through a combination of diet, weight loss to an acceptable body composition, aerobic exercise, and medication.

Disease Distribution of 430,000 Deaths Attributed Annually to Cigarette Smoking

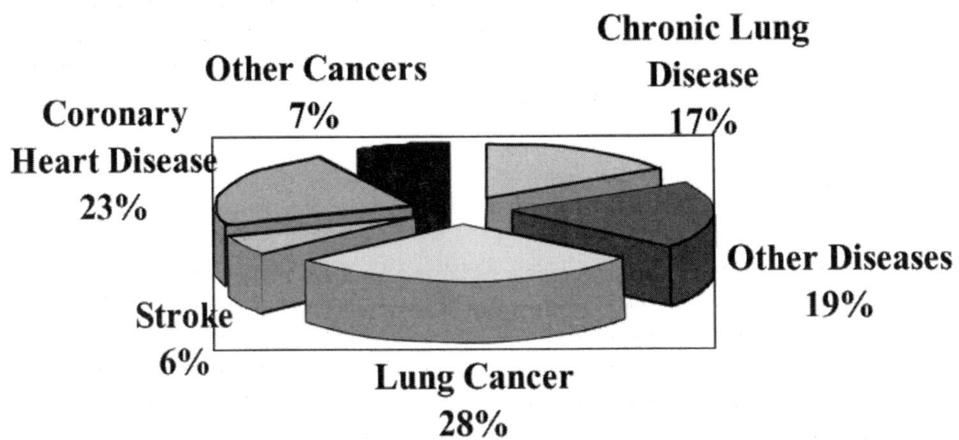

Smoking. Smoking is estimated to cause 430,000 premature deaths each year (see Figure 4.3) including 87 % of all deaths due to lung cancer, 30 % of all cancer deaths, and 21 % of deaths due to heart disease. Additionally, smoking leads to chronic bronchitis, emphysema, and, if done during pregnancy, low birth weights. While fewer persons are smoking each year, currently 29 % of all Americans 12 years and older smoke.

Body Fat. As excess body fat increases so does the incidence of cardiovascular disease, Type II diabetes, hypertension, various cancers, and osteoarthritis. With as little as 20 pounds of excess body fat, your chances of a heart attack triple. On the other hand, very low body fat also leads to a variety of disorders. These consist of abnormal physiological functioning leading to organ malfunctioning, osteoporosis, psychological problems, premature death, and menstrual irregularities for a female. Healthy body composition is 5 to 15 % body fat for males and 15 to 25 % for females.

Poor Physical Fitness. Recently, the American Heart Association included lack of physical activity as one of the primary risk factors of cardiovascular disease. There is a 2.24 times increased risk of early death for those with any level of body fatness as compared to a fit individual of similar body composition. In addition, regular physical activity reduces the risk of developing hypertension, Type II diabetes, and colon cancer. A recent study of 17,000 Harvard alumni revealed that those who regularly exercised at least three hours per week suffered 64 % fewer heart attacks than those who did not exercise. Also, of those who had a heart attack, 49 % who did not exercise regularly died within four weeks, whereas only 17 % died of those with a good history of physical activity.

TABLE 4.1

Blood Cholesterol Level Guidelines

	DESIRABLE	BORDERLINE	HIGH
• **Total Cholesterol**	< 200 mg/dl	200 - 239 mg/dl	> 239 mg/dl
• **LDL Cholesterol**	< 130 mg/dl	130 - 159 mg/dl	> 160 mg/dl
• **HDL Cholesterol**	> 44 mg/dl	35 - 44 mg/dl	< 35 mg/dl

Elevated Blood Fats.   A number of fats (i.e. lipids) are found in the blood and are essential for normal physiological functioning. These blood lipids include cholesterol, phospholipids, triglycerides, and fatty acids. Research has shown that high blood levels of cholesterol (see Table 4.1) are related to atherosclerosis. Cholesterol is an essential component for the formation of cell membranes, vitamin D, and sex hormones. The adult body synthesizes enough cholesterol to meet its daily needs, however infants and children require additional cholesterol through the diet. Consumption of animal tissue (e.g. eggs, meat, and cheese) adds cholesterol to the body. In order for cholesterol to be transported through the blood, the liver encases it with a protein and forms a lipoprotein.

There are two main types of cholesterol carrying lipoproteins: 1) low-density (LDL) and 2) high-density (HDL). LDL's are the primary transporters of cholesterol and normally carry between 60 and 80 % of the total blood cholesterol. These lipoproteins are known as "bad" cholesterol because they have the greatest affinity for the cells of the arterial walls where the cholesterol is deposited. These depositions begin the early stages of atherosclerosis. Aerobic exercise, body fat, and diet affect LDL concentration.

Good cholesterol or HDL protects against cardiovascular disease by acting like a scavenger. HDL removes cholesterol from the arterial walls and transports it back to the liver for conversion into bile and excreted through the intestinal tract. Regular aerobic exercise increases HDL levels, whereas smoking, using steroids, and being diabetic decrease the levels.

Emotional stress.   Stress that causes tension within the body can lead to a number of physical ailments including heart disease. Persons under high emotional stress have been found to have elevated blood cholesterol levels, elevated blood pressure, and increased incidence of heart disease.

Diabetes.   Diabetes is a disorder in which the body has chronically elevated blood sugar (glucose) levels. This condition occurs because the body is unable to transfer glucose from the blood into the cells where it can be used for energy. Diabetes is the third leading cause of death by disease in the United States. It can cause several diseases including cardiovascular, kidney, and peripheral vascular that lead to blindness and limb amputation.

There are two types of diabetes. Type I is insulin-dependent and occurs primarily in young individuals. This type requires daily injections of insulin to maintain normal blood glucose levels and is genetically related. Type II diabetes is non-insulin dependent and approximately 90 % of the 12 million diabetic Americans have this type. The cells in the type II diabetic's body have a resistance to insulin and are related to obesity. Treatment includes diet and exercise to control blood glucose levels and reduce body fat. Diabetics have double the heart-disease problems of nondiabetics. The life expectancy of a 35 year old untreated diabetic is 63 years, compared to the normal life expectancy of 76 years.

Aerobic Exercise

Aerobic exercise is a term used to describe a type of activity that derives its energy from aerobic metabolism. Exercising muscles requires energy to contract and move the body. For muscles to contract repeatedly for an extended period of time, oxygen is required to maintain the availability of energy for the working muscles. This need for oxygen must be met by the cardiorespiratory system or the contracting muscles will come to a stop. An exercise is considered aerobic if it uses large muscle groups in a repetitive, rhythmic motion for over three minutes. For a list of typical aerobic activities refer to Table 4.2.

TABLE 4.2

Typical Aerobic Activities for Cardiorespiratory Fitness

• Aerobic Exercise Machines	• Rope Jumping
• Bench-Stepping	• Rowing
• Bicycling	• Running
• Cross Country Skiing	• Stair Climbing
• Exercise to Music	• Swimming
• In Line Skating	• Walking

By improving the cardiorespiratory system, the risk of developing cardiovascular disease decreases. For an aerobic exercise to benefit the cardiorespiratory system, it must follow the overload principle (Chapter 2). This principle has three components: 1) frequency, 2) intensity, and 3) time.

1. Frequency

The minimum frequency of performing an aerobic exercise to produce an improvement in the cardiorespiratory system is three days per week. As the number of days per week of aerobic activity increases, so does the benefit. However for health fitness, no more than five days per week is recommended. Aerobically exercising six or seven days per week only minimally increases health fitness benefits but causes a greater chance of becoming injured.

2. Intensity

Exercise intensity per aerobic session must be in the Heart Rate Training Zone for positive adaptations to occur in the cardiorespiratory system. The training zone falls between 60 to 90 % of your maximum heart rate (see Figure 4.4). Maximum heart rate can be determined by performing a maximum exercise test in a laboratory setting or can be predicted by subtracting one's age from 220. However, when predicting maximum heart rate there is an error of plus or minus 10 heartbeats per minute. The

FIGURE 4.4

Heart Rate Training Zone (60–90%) for Age-Predicted Maximum Heart Rate.

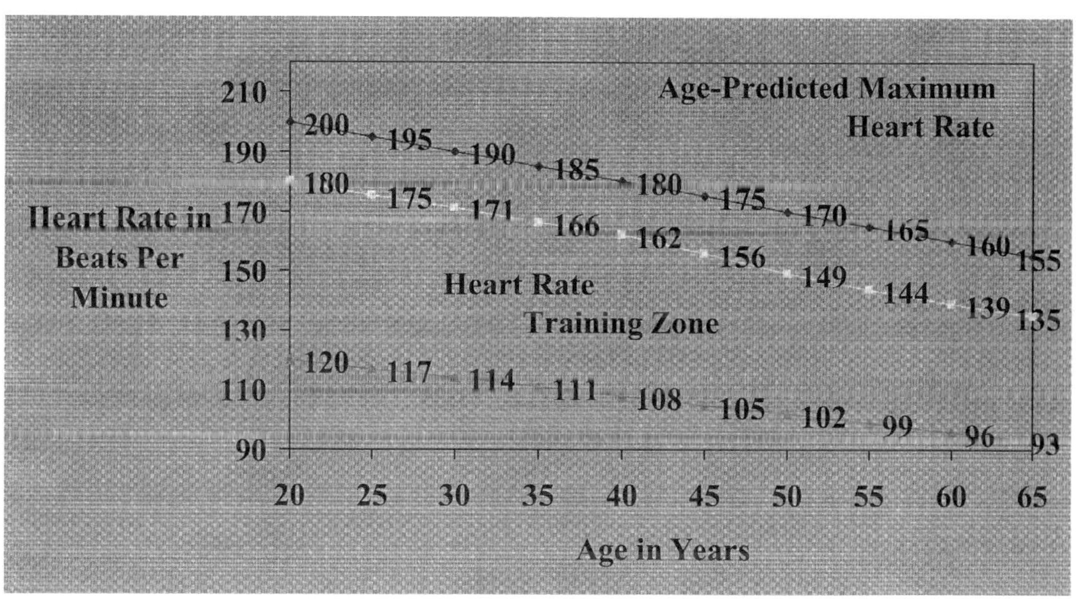

higher the intensity (heart rate), the greater the improvements in the cardiorespiratory system. However for health fitness, the intensity should be maintained between 70 to 80 % of maximal heart rate. Minimal adaptations occur at 60 %, but on the other hand, greater than 80 % increases the chance of injuries.

Measuring Heart Rate.   An easy method to measure heart rate is by feeling the pulsing blood as the left ventricle pushes it into the aortic artery. To feel your pulse, use the index and middle fingers pressed gently against the artery. The thumb should not be used as it has a pulse of its own and may cause an incorrect pulse count. The two most common arteries used for taking the pulse during exercise are the 1) carotid and 2) radial.

The carotid artery is located on either side of the throat (larynx). It lies in the valley between the Adam's apple and the neck muscle. This artery is usually the easiest to locate quickly, however there are special receptors located in the walls of the carotid that are responsible for detecting changes in blood pressure. By pressing too firmly while taking the pulse, these baroreceptors will send a message to the heart to slow down its contractions. This slowing of the heart rate may cause dizziness or faintness and will cause an incorrect heart rate reading.

The radial artery is located on the palm-side of the wrist. This artery can be felt on the outside of the forearm muscle tendons running through the wrist located on the thumb side. With practice the radial artery will be easy to locate, and at this site there are no baroreceptors to slow the heart rate.

To determine resting heart rate per minute count your pulse rate for 60 seconds. During exercise take a 10-second pulse count and multiply this number by six for your exercising heart rate per minute. To keep the exercise heart rate at the appropriate training zone, the pulse rate should be checked every five to 10 minutes. If there is difficulty in taking the pulse during the aerobic activity (e.g. swimming), stop and take the 10-second reading immediately. When you stop exercising, the heart rate only stays at that level for approximately 15 seconds before it begins to decrease rapidly. Don't be tempted to take a six-second-pulse count because it is easier to multiply by 10 mentally. This shorter time span magnifies any counting error by a factor of 10 possibly causing the exercise intensity to be out of the training zone.

Another method for determining aerobic exercise intensity is by using the ratings of perceived exertion scale. This scale was developed by a Swedish physiologist named Gunnar Borg. He discovered a relationship between the perception of exercise intensity and heart rate (see Table 4.3). The Borg scale has word descriptions that represents how hard you perceive you are exercising matched up to numbers. A perceived exertion number between 12 and 18 corresponds to the heart rate training zone of 60 to 90 % with 14 equaling 70 %. This procedure is not as accurate as heart rate but is easier to use.

3. Time

The amount of time spent performing the aerobic activity determines the magnitude of benefits to the cardiorespiratory system. The minimum amount of time for during the aerobic exercise for adaptation to occur is 20 minutes. However less time is required when the intensity of the workout is high. The reverse is true for lower exercising heart rates, which need a longer duration to gain positive adaptations. Table 4.4 shows the approximate heart rate and time needed to gain similar cardiorespiratory improvement.

TABLE 4.3

Perceived Exertion of Aerobic Exercise Scale

•	6
• **Very, Very Light**	7
•	8
• **Very Light**	9
•	10
• **Fairly Light**	11
•	12
• **Somewhat Hard**	13
•	14
• **Hard**	15
•	16
• **Very Hard**	17
•	18
• **Very, Very Hard**	19

Reprint from Bork, 1982

TABLE 4.4

Aerobic Exercise Intensity and Duration for Similar Cardiorespiratory Improvements

INTENSITY % of Maximum Heart Rate	TIME SPENT EXERCISING	
	Minimum	Optimal
• 60 %	40 minutes	60 minutes
• 70 %	30 minutes	45 minutes
• 80 %	20 minutes	30 minutes
• 90 %	10 minutes	20 minutes

Mode of Aerobic Exercise

There are no differences in the cardiorespiratory benefits with different types of aerobic exercise, as long as the frequency, intensity, and time are similar. However, in following the Principle of Specificity (Chapter 2), the type of aerobic activity performed will cause specific adaptation to other bodily systems (i.e. muscular, skeletal, and nervous). Because of this, the magnitude of improvements to be measured from the conditioning process, the testing mode, must be the same as the aerobic exercise. If an individual bicycles for aerobic exercise to improve their cardiorespiratory system, they will demonstrate the greatest fitness gains by being tested riding a bike. Other testing modes (i.e. running, swimming, or walking) will not fully measure the cardiorespiratory gains that have taken place from the aerobic conditioning process.

When deciding which form of aerobic exercise to participate in, a few guidelines should be considered. In following the Principle of Individuality, no single mode of aerobic exercise is the best for everyone. Each method has strengths and weaknesses specific to that activity (see table 4.5). One important factor in choosing is to pick the exercise that will be most enjoyable and will be the easiest to incor-

TABLE 4.5

Strengths and Weaknesses of Common Modes of Aerobic Exercise

AEROBIC ACTIVITY	STRENGTHS	WEAKNESSES
Lap Swimming	Develops upper-body strength. Low stress to joints. Good for disabilities.	Must be a skilled swimmer. Pool is required. Does not develop legs.
Outdoor Cycling	Develops leg strength. Low stress to joints. Use as transportation.	Cost for purchase & upkeep. Little upper-body development. Weather conditions.
Running	Minimal cost. Convenient. Requires minimal skill.	Stressful to joints. Little upper-body development.
Walking	Minimal cost. Convenient. Low stress to joints.	Greater time per session. Little upper-body development.
Aerobic Machines (stair, cycle & elliptical)	Develops leg strength. Low stress to joints. Read or watch TV while performing.	Cost for purchase & upkeep. Only eccentric contractions. Storage.
Cross-Country & **Rowing Machines**	Develops both upper & lower body. Low stress to joints.	Cost for purchase & upkeep. Only eccentric contractions Storage.
Exercise to Music (aerobic & step)	Music and variety of movements. Social interaction.	Cost for classes. Mainly develops lower-body.

porate into ones lifestyle. Regular aerobic exercise requires a commitment and belief that the minimal time spent each week will increase your productivity, longevity, and quality of life. To increase enjoyment and limit excessive amounts of stress to individual body parts, it is best to perform several different aerobic activities during the week.

Progression of Aerobic Exercise Program

For the body to continue improving cardiorespiratory fitness, the overload must increase periodically until the desired fitness level is reached. Small progressive increases in the aerobic exercise program will allow the body to adapt to each higher workload without excessive stress causing injuries. The maximal amount of overload increase per week for achieving health fitness is 10 %. If a person is running 10 miles per week, then next week they should be able to run 11 miles follow by 12.1 miles the week after that. If the same individual wants to increase the intensity instead of duration, then they should increase the heart rate for one workout per week by 10 %.

Benefits of Regular Aerobic exercise

If the aerobic conditioning stimulus follows the Overload Principle, the majority of the physiological adaptations occur independent of sex and age. These improvements will develop regardless of the mode of aerobic activity. Some of the following changes take place after only one exercise session, however most of these require many months of regular aerobic exercise to be fully acquired.

Metabolic Changes

The skeletal muscles used during the aerobic exercise will become more efficient in deriving greater amounts of energy from aerobic metabolism. These conditioned muscles will develop larger and more numerous mitochondria (i.e. where aerobic metabolism occurs) and increased aerobic system enzymes causing improved extraction of oxygen from the blood. In addition, the conditioned muscles have a greater capacity to mobilize, deliver, and utilize fats for energy that will also aid in maintaining cellular integrity.

Cardiorespiratory

The heart increases in weight and volume with long-term aerobic conditioning. The left ventricular cavity enlarges and the myocardium becomes thicker and stronger. These changes cause an increase in the stroke volume during both rest and exercise.

Because of the greater stroke volume, the heart rate decreases at rest and while performing submaximal activity.

Aerobic conditioning reduces both systolic and diastolic blood pressure at rest and during submaximal activity. This changes most noticeable in hypertensive individuals. Plasma volume increases and blood flow to working muscles becomes greater. Better distribution of blood also occurs which helps cool the body in a hot environment.

Other Changes

Regular aerobic activity leads to a reduction in body fat percentage. Also there are psychological benefits that include a reduction in anxiety and depression. In addition, aerobic exercise improves mood, self-esteem, self-concept, and general perception of personal worth. Finally, exercise performance increases with the conditioning activity causing faster times and quicker recovery from exercise.

Reversibility Principle

When an individual stops participating in an aerobic exercise program, deconditioning starts to occur. After 10 to 14 days of inactivity, there is a significant reduction in both metabolic and exercise capacity. Also, most of the conditioning adaptations are lost within several months. The rate of decrease is dependent on the conditioned level prior to stopping. However, even highly trained athletes from years of intense aerobic training will loose most of their adaptations within six month of becoming sedentary.

Summary

Risk factors for developing cardiovascular disease that are uncontrollable include heredity, gender, and age. Controllable risk factors include high blood pressure, smoking, obesity, lack of aerobic exercise, elevated blood fats, stress, and diabetes. Regular aerobic exercise can decrease the incidence of cardiovascular disease by improving the cardiorespiratory system. For an aerobic exercise to cause positive adaptations the Overload Principle must be followed. This includes performing the aerobic activity three to four days per week (frequency), at 60 to 90 % of age-adjusted maximal heart rate (intensity), and for 20 to 60 minutes per session (time). All types of aerobic exercises produce similar results, however these occur only in the systems and muscles used for the activity. Typical adaptation to an aerobic conditioning program are lower resting and exercising heart rates, a stronger and more efficient heart, increased oxygen utilization, and decreased blood pressure. Once the aerobic conditioning program is stopped the acquired adaptations start to be lost in 10 to 14 days.

Questions

1. What condition is the cause of 95 % of heart attacks and stroke?

2. Discuss the process of atherosclerosis.

3. List the uncontrollable and controllable risk factors that contribute to cardiovascular diseases.

4. What are the initial symptoms of an individual suffering from a heart attack?

5. What percentage of Americans has high blood pressure?

6. Define anaerobic exercise.

7. What is the Overload Principle requirement for aerobic exercise to benefit the cardiorespiratory system?

8. Which two arterial locations are recommended for counting one's heart rate?

9. List the strengths and weaknesses of swimming, cycling, running and aerobics to music.

10. List the cardiorespiratory adaptations from regular aerobic exercise.

5

The Skeletal Muscular System and Strengthening Exercises

Learning Objectives

This chapter explains the basic anatomy of the muscular-skeletal system and the importance of muscular endurance and strength. It also gives specific guidelines that should be followed in order to develop the muscular system. Upon completion of this chapter you should be able to:

1. Describe the structure of skeletal muscle.

2. List the characteristics for the three main muscle fiber types.

3. Describe specific guidelines for developing muscular endurance and strength.

4. Explain the Overload Principle as it pertains to strength training.

5. Demonstrate the three types of muscle contractions.

6. Name five adaptations to strength training.

Key Terms

Anaerobic Metabolism

Concentric Contraction

Delayed Onset of Muscle Soreness (DOMS)

Dynamic Strength Training

Eccentric Contraction

Hypertrophy

Isokinetic

Isometric Contraction

Isotonic

Motor Neuron

Motor Unit

Muscular Endurance

Muscular Strength

Repetition Maximum (RM)

Resistance

Set

Static Strength Training

Strength Training

Tendons

Type I Fibers

Type II Fibers

The Skeletal Muscular System

Fitness of the skeletal muscles of the body is of utmost importance. According to the American College of Sports Medicine, strength training that is adequate to develop and maintain muscle should be part of a well-rounded fitness program for adults. Human movement depends on muscles generating forces on the skeletal system causing the bones to move about their joint axes. However, skeletal muscles are also responsible for holding fixed body positions such as sitting or standing. In addition, muscles generate a large amount of heat (e.g. exercising and shivering) to help maintain body temperature in a cold environment.

The body has more than 660 skeletal muscles, which account for about 45 % of the male's body weight and 35 % of the female's weight. Approximately 75 % of skeletal muscle is water, 20 % is protein, and five percent is composed of various minerals, enzymes, fats, and carbohydrates. The functional unit of the muscle is the muscle cell called the fiber. Each muscle is made up of thousands of individual muscle fibers, with each surrounded by connective tissue. The connective tissue comes together at the end of the muscle to form a tendon, which is attached to a bone.

Human skeletal muscle is not made up of identical fibers with similar contractile and metabolic characteristics. Rather each muscle is composed of several different fiber types. The three main skeletal muscle fiber types are 1) slow twitch (Type 1), 2) fast twitch (Type IIb), and 3) intermediate (Type IIa).

Type I muscle fibers possess the potential for great aerobic metabolism. These fibers are well suited for aerobic activities and are resistant to fatigue. However, their contractile rate is slow and their force-generating capacity is low. Type I fibers are recruited first by the body for all types of activity.

Type IIb fibers have a great anaerobic (this concept will be further explained later in this chapter) capability. This fiber can generate 10 times the force during contraction as that of Type I fibers at five times the speed. These Type IIb fibers are used when the body needs to apply maximal force such as when sprinting, jumping, or serving a tennis ball.

Type IIa fibers are called "intermediate" because their characteristics are similar to both Type I and IIb fibers. Type IIa fibers can contract very rapidly and they have the capacity for both aerobic and anaerobic metabolism. These skeletal muscle fibers have the greatest potential for adaptation to the overload placed upon them.

The form of activity determines the specific kind of adaptation that will occur in each muscle fiber type. Through conditioning, all fibers in the exercising muscles will develop their existing characteristic potential based upon the type of overload. Each individual's skeletal muscle fiber type composition is determined by genetics and cannot be changed. The normal fiber type ratio for 95 % of the population is 50 % Type I (slow twitch) and 50 % Type II (fast twitch). This ratio averages out to a range from 60 % Type I and 40 % Type II to 40 % Type I and 60 % Type II. Elite athletes of power/strength or endurance events makeup part of the remaining five percent of the population who have less than 40 % or greater than 60 % Type I fibers, respectively. However, the majority of athletic events require use of both Type I and II fibers to be successful (see Figure 5. 1).

Skeletal Muscle Type I (Slow Twitch) Fiber Composition of Elite Athletes in Various Sports and Conditioned College Students

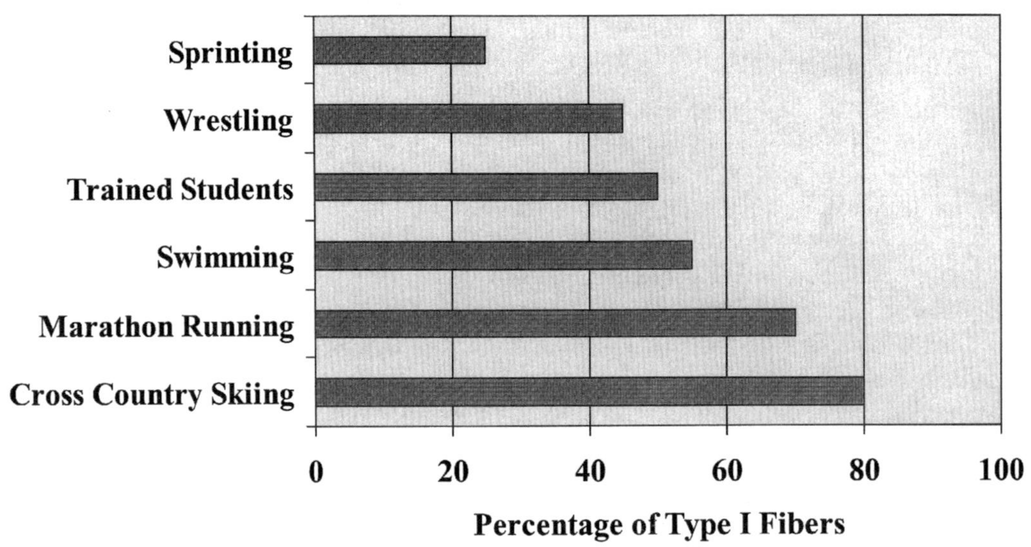

Types of Muscle Contractions

When a person is ready to move, a motor program is engaged and communicated to the appropriate skeletal muscle fibers via the nerve fibers called motor neurons. One motor neuron may attach (innervate) to just a few muscle fibers as in the fingers or several thousand as in the thigh. The motor neuron and all of the fibers it innervates is called a motor unit. All fibers in a motor unit are the same type and when stimulated, all will contract. The body regulates the strength of muscle contraction by recruiting the appropriate number of motor units. The greater the requirement of strength, the greater the number of motor units activated. There are three types of skeletal muscle contractions: 1) isometric, 2) concentric, and 3) eccentric.

1. **Isometric.** The term isometric is derived from two Greek words: 1) "*iso*" meaning the same and 2) "*metric*" meaning measure. An isometric muscle contraction occurs when the muscle fibers generate a force to shorten the muscle's length, but they are unable to overcome the external resistance. As a result, no movement occurs in the joint or muscle. Nevertheless, the muscle fibers are still creating a tremendous amount of force. An example of an action causing an isometric contraction is pushing against a wall.

2. **Concentric**. The most common type of muscle contraction is called concentric. This action occurs when the muscle fibers contract forcefully enough to overcome any external resistance and actually shortens the muscle's length. When the muscle shortens, it pulls the attached bones toward each other causing movement in the joint. An example of an action causing a concentric contraction is picking up a book.

3. **Eccentric.** When the external resistance is greater than the force generated by the muscle, the muscle fibers lengthen while tension is developed. This type of muscle contraction is called eccentric and acts as a brake to control the speed of movement caused by a force. This type of contraction causes the greatest amount of muscle soreness if the muscle is unaccustomed to the amount of force.

Modes of Strength Training

Various strength training methods are used to develop muscular endurance and strength. Each type has specific benefits as well as limitations. However, all resistant type activity comes under one of the two strength-training categories: 1) static and 2) dynamic (see Figure 5.2).

1. **Static Strength Training**. With static strength training the muscle generates a force to contract, but there is no change in muscle length. This type of resistance training is generally called isometric and is performed against an immovable object (e.g. wall or doorframe). The

FIGURE 5.2

Strength Training Modes, Methods, and Devices

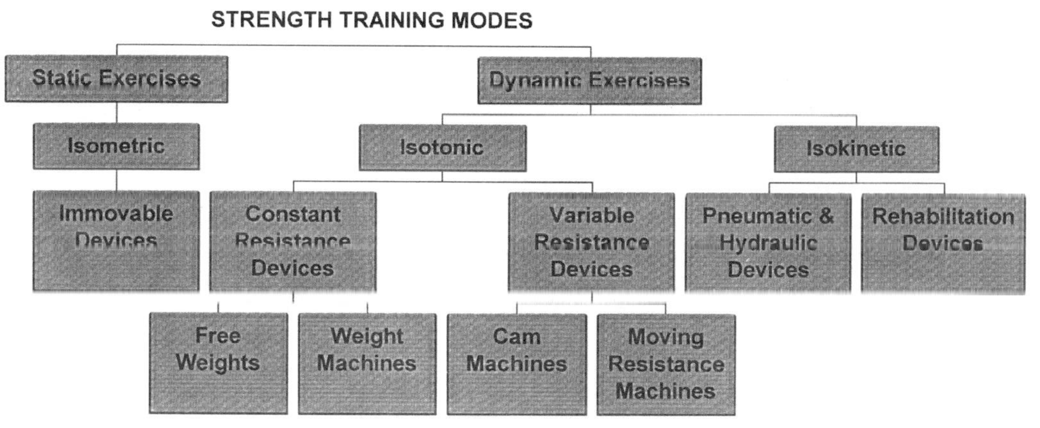

big advantage of isometric exercises is the cost because no specialized equipment is needed. The disadvantage is that muscle development from the overload only occurs at the specific angle of training. In addition, the straining nature of isometric exercises may lead to the Valsalva maneuver (chapter 3) which should be avoided by individuals with cardiovascular problems.

2. **Dynamic Strength Training**. Strength training involving muscle and joint movements is called dynamic and can be further subdivided into isotonic and isokinetic. Isotonic is derived from two Greek words: 1) "*iso*" meaning the same and 2) "*tonus*" meaning tension. Isotonic training uses several different methods, but all have resistance that remains constant throughout the range of motion. Typical isotonic devises are free weights (i.e. barbells and dumbbells) and weight machines. All isotonic devices have an inherent limitation (see Table 5.1), that throughout the lifting range of motion there are different positions where the muscles are not overloaded as much as other positions. To overcome this limitation, weight machines have been designed to vary the resistance throughout the lifting motion. This is attempted by having the mechanical levers move during the lift and try to increase torque in the body position where it is the strongest. At best, these devises do alter the amount of torque causing some increase in muscle overload in certain positions, however the weight being lifted does not change.

TABLE 5.1

Comparison fo Various Strength Training Devices

DEVICES	ADVANTAGES	DISADVANTAGES
• **Free Weights**	Good transfer to sport skills. Ease of Progression.	Need a spotter. Lifting technique required.
• **Cam and Weight Machines**	No spotter required. Little skill needed. Ease of performance.	Cost of equipment. Large weight increments. Not sport specific.
• **Hydraulic Machines**	No spotter required. Variable resistance. Ease of performance.	Cost of Machines. No eccentric contractions. Little transfer to sports skills.
• **Elastic Bands And Tubing**	Minimal cost. Use anywhere. Good for rehabilitation.	Poor strength development.

Isokinetic is derived from two Greek words: 1) *"iso"* and 2) *"kineo"* meaning to move. An isokinetic strength training method involves muscular contraction through a full range of motion performed at a constant velocity. The velocity of movement is the resistance factor instead of weight used in other dynamic training methods. Any force applied against the equipment results in an equal reaction force making it possible for the muscles to exert a maximal contraction throughout the full range of motion. These devises lend themselves for use in the rehabilitation of muscle and joint injuries. Since there is no active external forces being applied against the movement, injured or surgically repaired limbs can be strengthened safely.

Anaerobic Metabolism

Unlike aerobic metabolism (chapter 5), energy released through the anaerobic system does not require oxygen. The anaerobic system can generate large quantities of energy very rapidly, however this is only possible for a very limited time (less than two minutes). A byproduct of anaerobic metabolism is lactic acid, and accumulation of this product in the muscles interferes with energy production and muscle contraction. The main fuel source for the anaerobic system comes from carbohydrates (glucose), thus fats are not used during tense exercise.

Activities that primarily use anaerobic metabolism are high intensity but short in duration. Examples of these are strength training, sprinting, jumping, and playing intermittent type sports such as football, baseball, and basketball. Some aerobic benefits are gained from intermittent activities if performed with limited rest for an extended time period. However, much greater aerobic benefits will be acquired if aerobic activity is performed for that same amount of time.

Principles for Training Muscles

The four principles of physical conditioning (chapter 2) hold true for strength training just as they did for aerobic exercise. By incorporating these principles when designing a strength training program, health fitness goals may be more readily achieved. As little as 60 minutes per week of strength training is all that is needed for a successful program.

1. Overload Principle

The Overload Principle consisting of frequency, intensity, and time components must be followed for the skeletal muscles to adapt to strength training. In addition, skeletal muscles require a minimum of 48 hours rest between strength training

Continuum for Strength Training

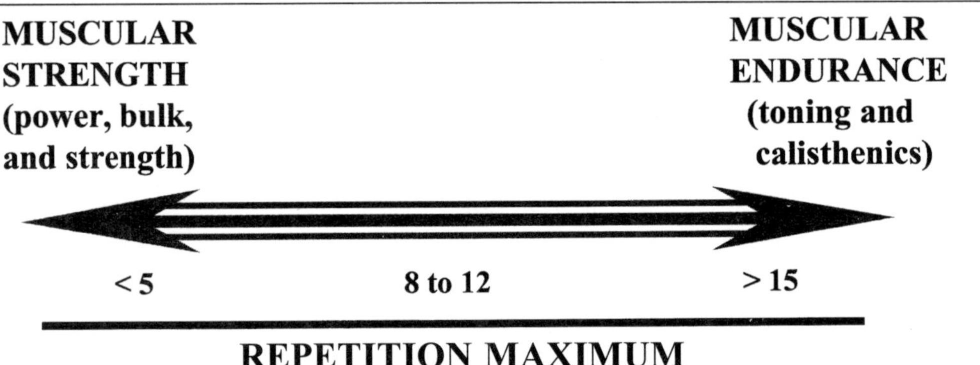

sessions for cellular repair. If the training workouts consist of just muscular endurance (e.g. sit-ups, pushup, and pull-ups), only 24 hours is needed.

Frequency of strength training refers to the number of sets (i.e. muscular contractions repeated to failure) performed for a certain exercise. Research has shown that one set per muscle group will cause an increase in strength. As the number of sets increase, so does muscular strength. However, after three sets the law of diminishing return takes over, and the extra amount of time spent does not equal the added increase in strength.

The intensity of strength training depends on the amount of resistance the muscles must generate against the force. This is typically the weight lifted, but some strength training devices use elastic bands or hydraulic systems. Intensity is based either on the maximal amount of repetitions (i.e. repetition maximum or RM) a weight can be lifted before muscle fatigue set-in or percentage of the maximal amount of weight that can be lifted one time (% of 1 RM). The greater the resistance, the greater the improvement in strength.

Time or the duration of the strength training exercise refers to the number of repetitions per weight lifting set. Muscular endurance and strength are closely related but are separate components of physical fitness. By placing both on opposite ends of a continuum, their relationship to strength training repetitions can be understood (see Figure 5.3). As the maximal amount of repetitions performed increases, the exercise will improve muscular endurance. The opposite is true for muscular strength, however. For developing both muscular endurance and strength, which is required for health fitness and most sports, 8–12 RM should be utilized.

2. Principle of Specificity

The adaptations from strength training are specific to the type of overload imposed upon the muscular-skeletal system. The following is a list of guidelines for developing specific muscular function.

1. To develop muscular strength, perform 5 RM or less.

2. For muscular endurance or toning, lift 15 RM or greater.

3. To improve both muscular endurance and strength or for health fitness, perform all strength training exercises 8 to 12 RM.

4. For muscular bulk/body building, lift large volumes of weight per muscle group.

5. One strength training session per week will increase muscular strength, however two sessions are better, and three will show the most gains.

6. For muscular development for sports skills, the strengthening exercises should mimic the skill pattern and speed of contraction.

7. To develop muscular power, the lifting exercises must be performed explosively.

8. A novice starting a strength training program should develop muscular endurance first then progress to muscular strength.

9. During the lifting or straining phase exhaling should occur, and while lowering the weight inhaling should take place. **NEVER HOLD THE BREATH DURING STRENGTH TRAINING** (see Valsalva maneuver chapter 3).

Muscle Soreness.   Along with any increase in physical training, muscle soreness occurs. There are two types of muscle soreness: 1) acute and 2) delayed. Acute muscle soreness begins near the end of the training session or immediately after it stops. This type of soreness is due to ischemic conditions and a build up of lactic acid in the exercised muscles. This is only temporary and after a few hours the soreness will diminish.

The second type of muscle soreness occurs 24 to 48 hours after the unaccustomed workload and is called Delayed Onset of Muscle Soreness (DOMS). This soreness is much more intense and lasts for several days. The cause for DOMS is actual cellular damage and happens predominately from performing eccentric muscle contractions. After the muscle repairs this damage it becomes stronger and is able to handle this same workload without soreness or damage. This is the natural adaptation process from strength training.

Adaptations to Strength Training.   With weekly strength training sessions the body will adapt neurologically and physiologically. To what extent will be based upon the specific type of overload, the months of training, and muscle fiber composition. The following is a list of what can be expected from strength training:

1. The central nervous system increases activation, which improves synchronization of motor units and creates more efficient neural recruitment patterns.

2. Muscle fibers increase in size (hypertrophy) but not in number (hyperplasia).

3. Anaerobic metabolism increases.

4. Ligament and tendon strength increases.

5. Bone mineral content increases making the bones stronger and decreasing the risk of osteoporosis.

3. Principle of Reversibility

Once the weekly strength training sessions are discontinued, detraining begins, and the body reverts back to a pretraining state. This reversing of the gained muscular, skeletal, and neural adaptations occurs after 10 to 14 days of inactivity. However with two training sessions per week, most adaptations can be maintained. With even one strength training session per week, some gains will remain. Interestingly, once the training resumes, the previous strength trained state takes less time to achieve.

4. Principle of Individuality

The extent of muscular development from a weekly strength-training program is not only dependent on the type of overload but also the individual's gender, age, and genetics.

Regardless of gender, skeletal muscle can generate three to eight kilograms of force per square centimeter of cross sectional area. However, males posses a greater absolute strength due to their greater muscle mass. This occurs because males have much higher levels of testosterone, allowing greater muscular hypertrophy. Even after years of intense strength training, most females will be unable to develop large muscles.

The aging process causes a decrease in skeletal muscle motor units and muscle fiber atrophy leading to a reduction in muscle mass and strength. However, with an active lifestyle including regular strength training this inevitable decline is drastically slowed. In addition, recent studies clearly show that skeletal muscles of elderly individuals who have been sedentary for many decades can adapt to strength training and develop muscular endurance and strength, even into the ninth decade of life.

Genetic predisposition will determine the extent of skeletal muscle adaptation to high resistant strength training. The greater percentage of Type II fiber in the trained muscle, the greater the amount of hypertrophy, strength, and power. In addition, the higher the level of testosterone, the greater and faster the adaptation to strength training. This is true for both males and females, young or old.

Summary

Muscular endurance and strength are basic requirements for daily living as well as for sports performance. Skeletal muscles are made-up of three main fiber types with each having specific characteristics. When muscle are stimulated by the nervous system to contract, they generate a mechanical force by pulling on the bones. To develop the skeletal muscles, an overload must be applied. The type and amount of overload will determine the specific adaptations to the muscles regardless of gender, age, or genetics. By manipulating the amount of sets, repetitions, and resistance, muscular endurance and/or strength can be acquired. There are several methods in achieving the desired strength training goal with a variety of devices to utilize. Strength training requires the skeletal muscles to utilize anaerobic metabolism, causing specific adaptation to this system. If weekly training sessions stop, detraining occurs and acquired benefits are lost. It is never too late to begin a strength training program and reap its many benefits.

Questions

1. Muscles account for approximately what percent of total body weight of males and females?

2. What are the recommended frequency, intensity, and time/duration for developing both muscular endurance and strength?

3. Define strength and muscular endurance.

4. List the three major muscle fiber types and two characteristics of each.

5. Discuss gender differences with regard to absolute strength and muscle hypertrophy.

6. Explain the three types of muscle contractions and give an example of each.

7. List several adaptations to strength training.

8. Define DOMS and tell what causes it to occur.

9. Name the three methods of strength training and an advantage of each.

10. How quickly does detraining start to occur?

6

Starting an Exercise Program for Health Fitness

Learning Objectives

This chapter explains the importance of properly preparing the muscles and joints for physical activity. Also how to warm-up, cool-down, and increase flexibility for health fitness. After reading this chapter you should be able to:

1. Understand the importance of warming up the body before starting to exercise.

2. Explain the differences between controlled stretching and flexibility training.

3. Describe the importance of proper hydration levels during exercise.

4. List the physiological adaptations from heat acclimatization.

5. Design a health fitness exercise program for yourself.

Key Terms

Acclimatization

Ballistic Stretching

Controlled Stretching

Cool-down

Dehydration

Flexibility Training

General Warm-up

Hyperhydration

Medical Clearance

Rehydration

Specific Warm-up

Static Stretching

Medical Clearance

Before a person begins any exercise program, they should consult with a physician to determine if they should have a complete physical examination. This is especially true for individuals with any disease or physical limitation (e.g. overweight or pregnant). According to the guidelines of the American College of Sports Medicine, healthy men under age 40 and healthy women under age 50 with no symptoms of cardiovascular disease and less than two risk factors (chapter 4) do not require medical evaluation before starting a vigorous exercise program (Laboratory 1).

Exercising Session

Each exercise session should contain four components: 1) warm-up, 2) exercise, 3) cool-down, and 4) flexibility training. By utilizing all four, the conditioning session will be more productive with a decrease chance of injuries.

1. Warm-up

Prior to any exercise activity a warm-up should be performed. There are several benefits that occur from a properly warmed body (see Table 6.1) including a less chance of injuries. A good warm-up consists of three parts: 1) general warm-up, 2) controlled stretching, and 3) specific warm-up.

TABLE 6.1

Physiological Consequences of a Proper Warm-up

- 1. Increased blood flow through the active muscles.

- 2. Facilitation of oxygen delivery to the muscles

- 3. Greater efficiency of movement.

- 4. Increased speed of muscle contraction.

- 5. Facilitation of nerve transmissions.

- 6. Improved myocardial blood flow.

- 7. Reduced myocardial workload at the start of exercising.

General Warm-up.   The general warm-up includes general body movements utilized to warm the body to a mild sweat or flushing of the skin and to slightly increase respiration. These movements are usually unrelated to the anticipated exercise activity or sporting event and are performed for two to three minutes. Examples of a general warm-up are calisthenics, walking, and jumping rope.

Controlled Stretching.   Mild stretching prior to the exercise activity helps to improve joint mobility and function. This type of light stretching is not designed to increase flexibility but to prepare the joints to be utilized during the workout. Each stretching activity requires slow-controlled movements with a three to four second hold at the point of discomfort and may be repeated up to three times. Figures 6.1 to 6.9 are examples of stretching exercises that can be utilized for most physical activities.

FIGURE 6.1

Shoulder Stretch Across the Chest. Grab your elbow and slowly pull it across to the opposite shoulder. Hold this position while relaxing the stretched muscles. Repeat using the other arm.

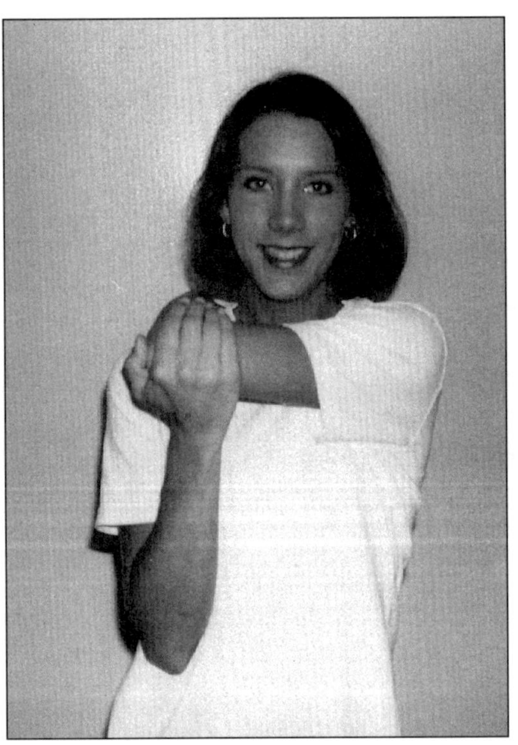

FIGURE 6.2

Shoulder Stretch Behind the Head. Raise an arm overhead and flex it next to the ear. Grab the elbow and slowly pull it behind your head. Hold this position while relaxing the stretched muscles. Repeat using the other arm.

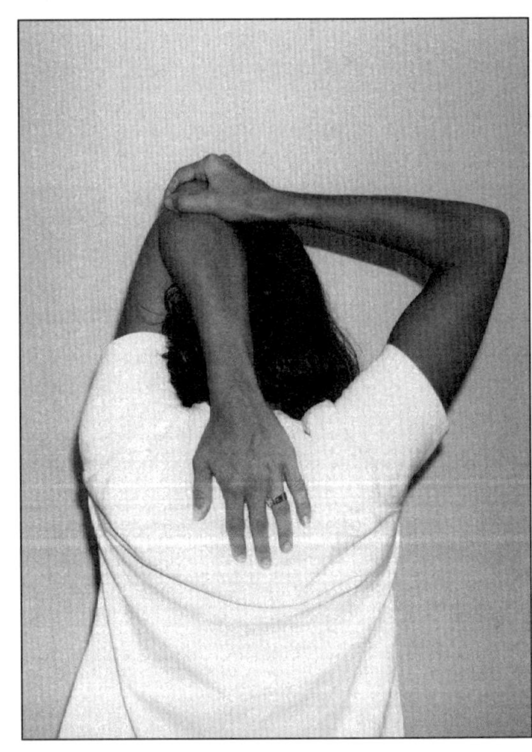

FIGURE 6.3

Shoulder Stretch to the Side. Grab a tall stationary object with on extended arm at shoulder height. Turn your body away from the arm, stretching the anterior shoulder. Hold this position while relaxing the stretched muscles. Repeat using the other arm.

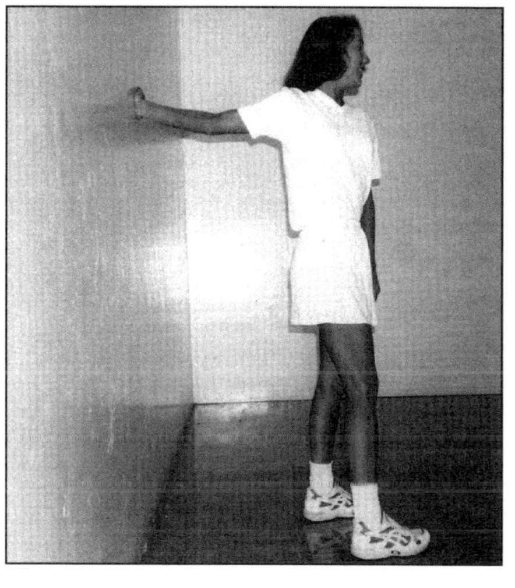

FIGURE 6.4

Trunk Stretch to the side. Stand with hands overhead and feet shoulder width apart. Grasp wrist with one hand and pull the arm down the side. Hold this position while relaxing the stretched muscles. Repeat to the other side.

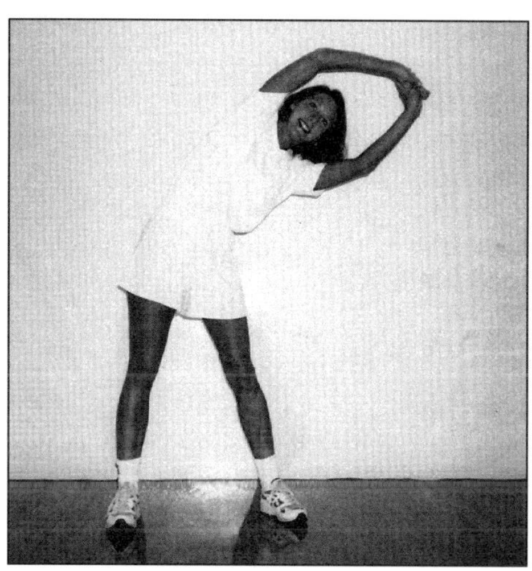

FIGURE 6.5

Thigh Stretch. Standing near an immovable object, rest one hand against it for balance and support. Flex the opposite side leg and grasp the foot pulling the foot toward the buttocks. Hold this position while relaxing the stretched muscles. Switch legs and repeat.

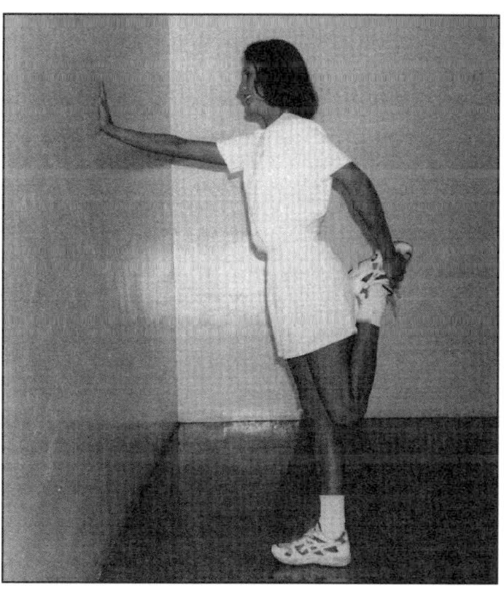

FIGURE 6.6

Achilles Stretch. Stand facing an immovable object and place hands upon it for balance. Extend one leg back while flexing the other. The extended leg's foot is flexed with the heel in contact with the floor and toes pointing forward. Hold this position while relaxing the stretched muscles. Switch legs and repeat.

FIGURE 6.7

Groin Stretch. Sit on the floor with knees flexed and soles of feet together. Grasp ankles and pull toward the groin while pressing down with the elbows on the knees. Hold this position while relaxing the stretched muscles.

FIGURE 6.8

Hip Stretch and Trunk Twist. Sit upright on the floor with one leg flexed across the other. Turn toward the bent leg and using the opposite side arm, press against the knee. Look and twist the trunk as far as possible. Hold and relax the stretched muscles. Repeat by switching the legs and twisting in the opposite direction.

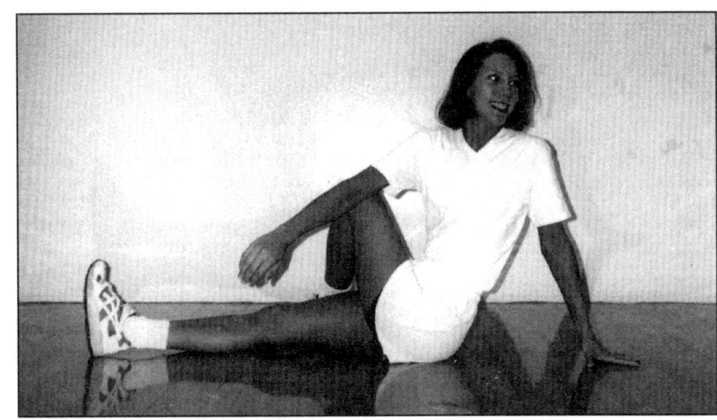

FIGURE 6.9

Hamstring Stretch. Sit on the floor with one leg flexed and pulled to the groin. With the other leg extended, bend at the waist and lower the chest toward the thigh. Hold this position while relaxing the stretched muscles. Switch leg positions and repeat.

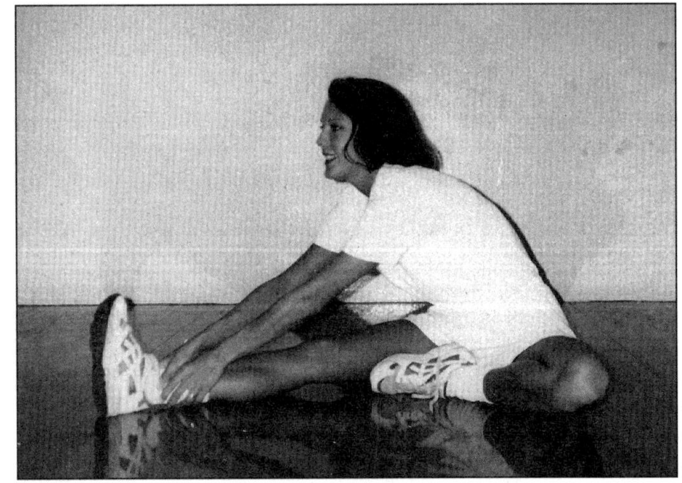

Specific Warm-up. The type of physical activity being performed during the workout will determine the kind of exercises utilized for the specific warm-up. The purpose of this final part of the warm-up is to allow the participant to become familiar with the conditioning, training, or playing environment. The specific warm-up for an aerobic workout would consist of performing the aerobic activity by starting at a low intensity and slowly working up to the scheduled heart rate training zone percentage (Chapter 4). If the exercise activity is going to be strength training, then each lifting exercise will be executed first using little resistance for 10 to 15 repetitions, followed by more intense lifting. For a sporting activity, the specific warm-up would provide a skill rehearsal for the actual activity for which the warm-up is being utilized. Examples include swinging a golf club, throwing a baseball, and warm-up volleying in tennis.

2. Exercise Activity

The largest portion of time spent during the workout session should be devoted to the exercise activity. This is the component for which the majority of physiological, neurological, and psychological adaptations occur if the Overload and Specificity Principles are followed. For health fitness, the exercise activity should consist of aerobic exercise (Chapter 4) and/or strength training (Chapter 5). If the physical activity is a sport then skill drills and/or game play will take place during this part.

Regardless of the type of exercise being performed, proper athlete attire specific for the physical activity is important to limit the risk of injuries. This includes footwear, undergarments, outerwear, and protective equipment. The price of such apparel may appear costly, however this is minimal compared to the cost for a visit to the emergency room and the associated physical discomfort.

3. Cool-down

The cool-down begins as soon as the exercise session has ended and is designed to bring the body back to near resting levels. A brief period of low-intensity activity such as light jogging, swimming, cycling, or walking will facilitate the body's return to the pre-exercise state. An active cool-down is important immediately after a workout otherwise the blood will pool in the exercised muscles, decreasing the volume of blood returning to the heart and causing a decrease in cardiac output (Chapter 3). If this occurs, blood flow to the brain is diminished causing lightheadedness, dizziness, and possible fainting. By actively contracting and relaxing skeletal muscles, blood flow is "milked" back to the heart maintaining cardiac output. Other benefits of an active cool-down are to facilitate removal of lactic acid, to replenish oxygen stores, and to help dissipate heat.

4. Flexibility Training

Stretching to improve or maintain the range of motion of the major joints in the body should be an important component of every exercise session. As a person ages, the connective tissue surrounding each skeletal muscle and its fibers that form the tendon (Chapter 5) loses some of their elastic components. This causes a decrease in muscle-tendon flexibility and joint range of motion leading to a greater risk of injuries. A sedentary lifestyle also causes a decrease in resilience that compounds this aging process. Injuries to the muscle-tendon unit will likewise cause a decrease in flexibility due to the scar tissue formed during the healing process. However, if controlled stretching is performed as the body is repairing this injury, the joint range of motion will be maintained and the healed site will be less susceptible to future injury.

A planned, deliberate, and regular flexibility training program will progressively increase the pliability of the muscle-tendon unit leading to a decreased chance of muscle, tendon, and joint injuries. Beginning a flexibility training program at any age will cause improved functioning in the muscle-tendon unit and will increase its range of motion.

Flexibility training should begin towards the end of the cool-down while the muscles, tendons, and ligaments are warm and pliable. There are two types of stretching methods used for flexibility training. The first method is called ballistic stretching and consists of bouncing or rhythmic motions. This technique has the potential for causing injuries to the muscle-tendon unit due to large and uncontrollable amounts of momentum generated from the quick movements. Also the stretch reflex phenomenon causes an increase in muscular tension making it more difficult to stretch the muscle and defeating the purpose of flexibility training.

The second method of flexibility training is called static stretching and involves holding a position for a period of time. Static stretching uses slow and controlled movement to reach that held position. This position should be at the point of feeling some discomfort in the stretched muscles-tendon unit, and it should be held for 15 to 60 seconds. The longer the stretched position is held and the more times it is repeated the greater the improvements in flexibility. The exercises used for controlled stretching may also be used for flexibility training (see Figures 6.1 to 6.9).

Static stretching can be performed by oneself or with the help of a partner; however, partner stretching has the potential to improve more efficiently muscle-tendon flexibility and joint range of motion. There are two basic stretching techniques used with a partner, and these are designed to utilize the neuromuscular physiology. The first uses the effects of reciprocal innervation so that when a muscle (agonist) is contracted, the muscle performing the opposite movement (antagonist) will be induced to relax. As this relaxation occurs, the partner stretches this muscle to the point of discomfort and holds this position for 15 to 60 seconds.

The second method of partner stretching uses the concept of maximal contraction of a skeletal muscle leads to maximal relaxation in that same muscle. This technique requires the partner to resist the maximal contraction of the muscle soon to be stretched. After this maximal isometric contraction (Chapter 5), the partner stretches this relaxing muscle to the point of discomfort and holds this position for 15 to 60 seconds.

For maximum results from partner stretching, each exercise should be repeated up to five times. Disadvantages of this method include the requirement of a partner and the large amount of time it takes to complete each stretch. If partners are performing this technique on each other, the time doubles.

Hydration

Maintaining proper body hydration levels during an exercise session is critical for health, safety, and optimal performance. Dehydration leads to heat illnesses (i.e. muscle cramps, heat exhaustion, and heat stroke) and possible death. Normal hydration levels become more difficult to maintain as the exercise session increases and/or the exercising environment has high temperatures with or without high humidity. There are three distinct hydration periods for optimal body fluid levels: 1) pre-exercise, 2) during exercise, and 3) post-exercise.

1. Pre-Exercise Hydration

Proper hydration levels before starting the physical activity is necessary regardless of the length of the exercise session or environmental conditions (i.e. hot or cold). Any fluid deficit prior to exercise can potentially compromise thermoregulation during the session. This dehydrated state increases the cardiovascular strain and limits the body's ability to transfer heat from the contracting muscles to the skin surface where heat can be dissipated. This condition also causes a decrease in performance for both aerobic and anaerobic activities.

During the 24 hours period prior to the exercise session adequate fluids should be consumed to promote proper hydration. Starting about two hours before the exercise activity, approximately 500 milliliters or 17 ounces of water should be ingested. This fluid intake should be consumed gradually to elevate the hydration level above normal values causing a hyperhydrated state. If this pre-exercise water is consumed too quickly, the kidneys will slow their reabsorbtion rate leading to excretion of the excess ingested water in the urine.

2. Hydration During Exercise

During exercise humans typically drink insufficient amounts of fluid to negate water loss from sweat. Without adequate fluid replacement during physical activity, internal body temperature rises along with heart rate, and sweat rates decrease. To minimize the chance of developing heat illnesses, fluid intake should equal sweat rate. This approximates 150 milliliters (4 oz) to 350 milliliters (10 oz) every 15 to 20 minutes of exercise. The best fluid replacement is cool water for physical activities lasting under one hour. When exercising longer than one hour, a sport drink with less than 10 % carbohydrate concentration and less than seven grams of sodium per liter should be consumed. The carbohydrates (Chapter 8) will help maintain blood glucose levels and delay onset of fatigue.

3. Post-Exercise Hydration

Immediately after the completion of the exercise activity and during the cool-down period, rehydration should begin. Any type of fluid is acceptable as long as it does not contain caffeine or alcohol. Both of these drugs are diuretics and defeat the rehydration effort. If a person is a competitive athlete, the rehydration fluid should contain carbohydrates to help replenish the depleted skeletal muscle glycogen stores.

Exercising in the Heat or Cold

The human body must maintain internal (core) temperature within a very narrow range for normal physiological functions. Failure to do so will result in death. The body is very adept at regulating core temperature in varying environmental conditions during exercise. However, certain precautionary procedure should be followed to eliminate the potential for thermal illnesses.

Hot Environment

Exercising in a hot environment places additional stress on the body's cooling mechanisms. Skeletal muscles only use approximately 40 % of their generated energy to contract; the other 60 % is given off as heat. To dissipate this heat buildup, the body must dilate the arterioles leading to the skin capillaries allowing the blood to transport heat from the core to the skin. However, this means that there will be less blood flowing to the exercising muscles for its metabolic needs.

Another mechanism the body uses to remove excess body heat is through the evaporation of sweat. This is the body's major physiological defense against overheating. The water secreted from the sweat glands comes mainly from the blood causing a decrease in plasma volume and limiting the amount of blood available for skin capillaries to dissipate heat and for muscle metabolism. If exercising continues with the diminishing blood volume, core temperature will rise and heat illness will occur. This is precisely the reason for having proper hydration before and during exercise.

TABLE 6.2

Physiological Adaptations Due to Heat Acclimation

> Improved blood flow to the skin capillaries.

> Better distribution of blood flow.

> Sweating at a lower core temperature.

> Increased sweat output.

> Lower concentration of salt in the sweat.

> More effective sweat distribution over the skin surface.

Acclimatization to Heat. Physical activities that are non-taxing when conducted in a cool environment become difficult when performed in a hot one. However, repeated exposure to a hot environment while exercising causes the body to adapt, resulting in an improved capacity and less discomfort (see Table 6.2). Major acclimatization occurs during the first week of heat exposure and is completed after 10 days. The first few day of exercise in a hot environment should be of low intensity and last less than 30 minutes. Thereafter the exercise sessions can increase in time and intensity. Elderly individuals without a compromised cardiovascular system can adapt to a hot environment and regulate core temperature as easily as younger individuals.

High humidity and high environmental temperatures decrease the effectiveness of sweating. As relative humidity increases, less sweat evaporates to cool the body and there is a greater risk for developing heat illnesses (see Table 6.3). In addition, when individuals (fit or otherwise) exercise unacclimated to a hot and humid environment they are at higher risk. On days with high temperatures and humidity, outside exercise should be performed only in the morning or early evening. If exercise must be done during the day, it should be in an air-conditioned building.

Warm-weather clothing should be of lightweight material and be loose fitting to permit air circulation and sweat evaporation. Also light colored clothing is cooler since it reflects heat rays as opposed to dark colored clothing, which absorbs light rays.

TABLE 6.3

Heat-Stress Index

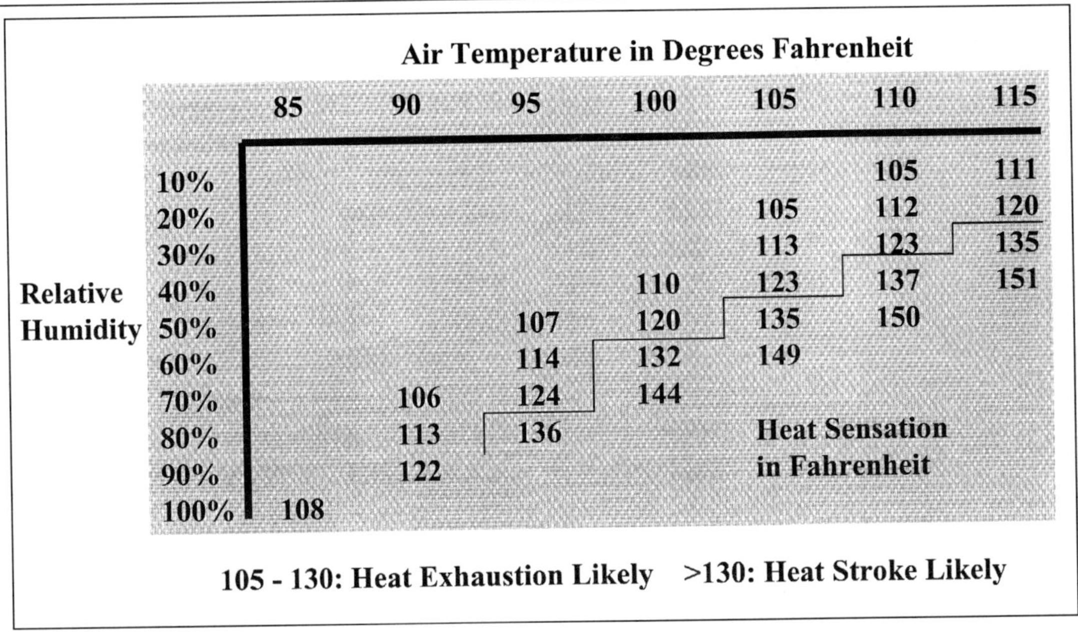

Cold Environment

Exercising in a cold environment places less stress on the body to maintain core temperature than a hot environment. The large amount of heat generated by exercising muscles can sustain a constant core temperature in air as cold as – 30 degrees C (-22 degrees F) without the need for heavy clothing. Hydration is still a concern in a cold environment; even though there is less sweating, greater amounts of water vapor are needed to humidify the inspired (Chapter 3) dry, cold air. Asthmatics should not exercise in a cold environment since the dry air may cause an attack.

Clothing for exercising in the cold should consist of several layers so that as the body warms layers can be removed. Also, the material against the skin should be effective in allowing water vapor to escape from the body's surface. Hats and gloves are important, as these coverings will keep the skin surface from freezing in a subfreezing climate.

Exercise Program Design for Health Fitness

When designing an exercise program for achieving health fitness one must first determine present fitness level after receiving medical clearance from a physician (Laboratory 1). There are several tests that will measure the participant's initial level on each of the five health fitness components (Chapter 2). The evaluating procedures and fitness rating scales are located in the laboratory sections of this book (Chapter 11).

1. Body composition evaluation, Laboratory 3.

2. Cardiorespiratory fitness evaluation, Laboratory 4.

3. Muscular endurance/strength evaluation, Laboratory 5.

4. Flexibility evaluation, Laboratory 6.

These fitness tests will assist the individual in discovering areas of health fitness that need more attention and others that only require maintenance. With the results from these laboratories, a health fitness exercise program can be designed that meets the individual's specific needs. To fit all the important components into a busy schedule takes careful planing and requires dedication and sacrifices. However, the benefits to be gained are great (Chapter 2). The exercise program design must follow the four Principles of Conditioning: 1) overload, 2) specificity, 3) individuality, and 4) reversibility. Tables 6.4–6.6 gives guidelines on designing a health fitness exercise program. Once the exercise program is started, it is important to progressively increase the overload until a fitness level of "Good" (tables 7.2, 11.4, 11.5 and 11.6) has been reached in all health fitness components.

TABLE 6.4

Starting Exercise Levels for Aerobic, Strength, and Flexibility Training Health Fitness Program, Based on Classification from Figures 6.10 and 6.11

EXERCISE STARTING LEVEL	AEROBIC TRAINING			STRENGTH TRAINING			FLEXIBILITY TRAINING		
	Freq.	Inten.	Time	Day/Wk	Sets	RM	Day/Wk	Reps	Time
I	3	50%	10min	2	1	20	7	3	25sec
II	3-4	60%	20min	2-3	2	15	5	3	20sec
III	3-5	70%	30min	2-3	2-3	8-12	3-4	2-3	15sec
IV	3-6	80%	40min	3	3	8-12	3	2	10sec

FIGURE 6.10

Determining Starting Level for an Aerobic Exercise Program Based on Body Fat Percentage and Maximal Oxygen Uptake from a 1.5 Mile Run/Walk

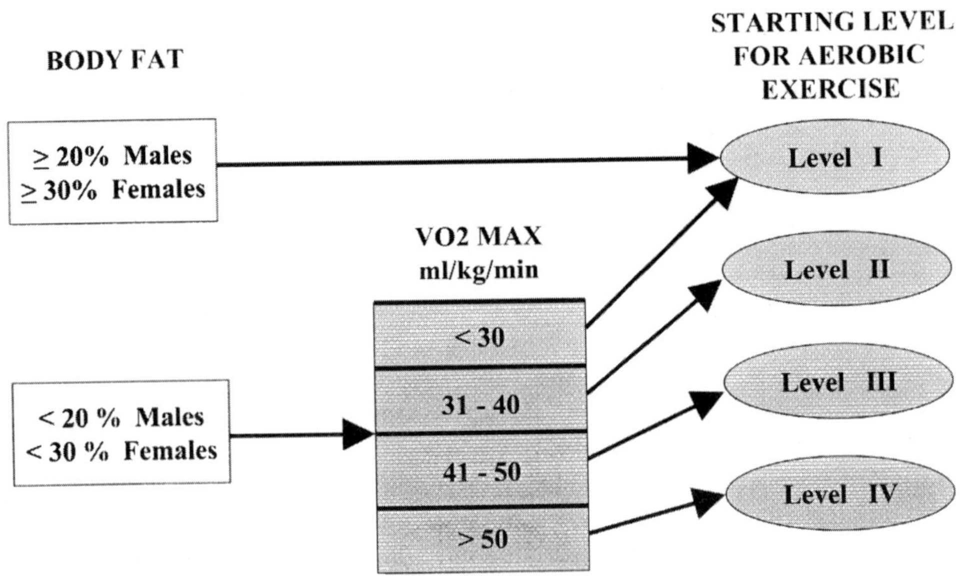

FIGURE 6.11

Determining Starting Levels for a Strength and a Flexibility Training Program Based on Three Muscular and Three Flexibility Tests

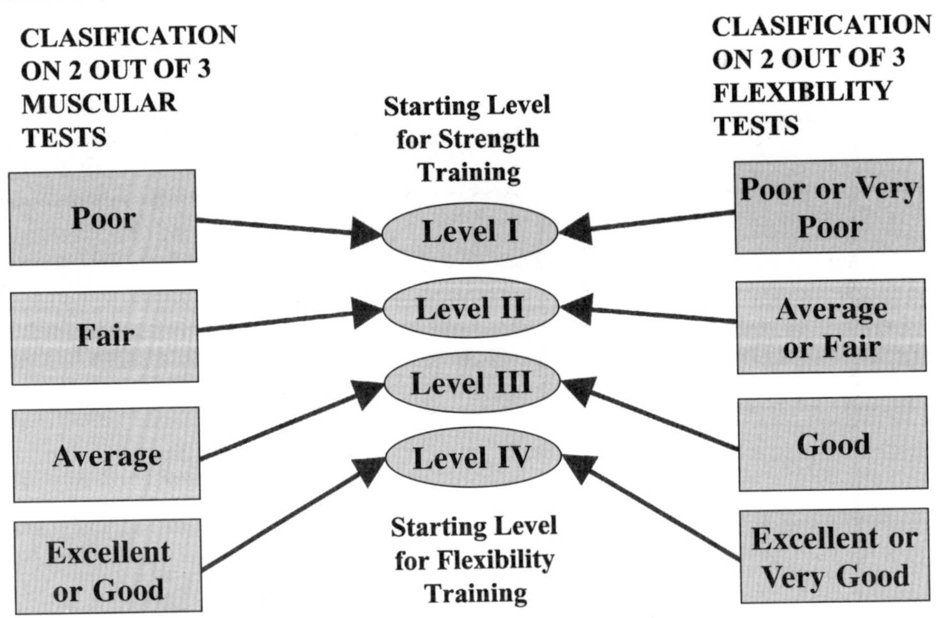

Summary

Before beginning an exercise program, medical clearance should be obtained from a physician. When designing an exercise program the first procedure is to determine one's present fitness level on the five health fitness components. With this information an exercise program can be designed to meet the individual's needs. Each exercise session should contain four components: 1) warm-up, 2) exercise activity, 3) cool-down, and 4) flexibility training. The warm-up consists of a general warm-up followed by controlled stretching and ending with a more specific warm-up. After the exercise session a cool-down should be performed to assist the body in returning to pre-exercise level. Flexibility training is designed to increase the range of motion in a joint and should be part of the cool-down. Static stretching is the best method to improve flexibility and can be performed by oneself or with a partner. When exercising it is critical to be properly hydrated to decrease the risk of developing heat related illnesses.

Questions

1. What is the importance of performing a good warm-up before exercising?

2. Explain the difference between controlled stretching and flexibility training.

3. Why is a cool-down period important after exercising?

4. Describe proper hydrating techniques that decrease the chance of heat illnesses.

5. How does the body adapt to a hot environment?

6. What physiological changes occur when the body acclimatizes to a hot environment?

7. When designing an exercise program for health fitness, what components should be included?

7

Body Composition and Health Fitness

<table>
<tr><td>

Learning Objectives

This chapter explains the importance of understanding body composition and energy expenditure for assisting one in achieving a healthy and fit body. Health fitness standards for percent body fat are presented. Caloric intake and caloric expenditure are examined as well as the benefits of having a physically active lifestyle. Upon completion of this chapter you should be able to:

1. Know the difference between body weight and body composition.

2. Distinguish between acceptable and unhealthy ranges for body fat for males and females.

3. Understand the technique for taking skinfold measures.

4. Know some of the health problems and diseases associated with being in an overfat condition.

5. Explain the difference between essential fat and storage fat.

6. Understand the energy balance equations and its relationship to weight control.

7. Describe three eating disorders and characteristics of each.

8. List the benefits of exercise in fat reduction.

</td><td>

Key Terms

Adipocyte

Anorexia Nervosa

Basal Metabolic Rate

Binge Eating Disorder

Bioelectrical Impedance

Body Mass Index

Bulimia Nervosa

Energy Balance Equations

Essential Fat

Hydrostatic Weighing

Lean Tissue

Overfat

Purging

Skinfold Measurement

Storage Fat

</td></tr>
</table>

Overfatness, Disease, and Related Costs

In 1990, the American population was 30 % overweight (overfat), and the resulting increase in diseases due to overfatness (see Table 7.1) threatened to breakdown an already overburdened health care system. This prompted the United States Department of Human Services, in conjunction with over 20 health and fitness organizations, to develop a set of health goals to be achieved by the year 2000. These goals were published under the title "Healthy People 2000: National Health Promotion and Disease Prevention Objectives". One of the goals was to reduce the prevalence of overfatness among American adults to no more than 20 % by the year 2000. Unfortunately by 1999, over fatness in the American adult population increased to 35 % putting over 80 million Americans at a higher risk for a variety of diseases.

Today the health care cost of overfat-related illnesses in the United States is over $56 billion annually (see Figure 7.1). This staggering number of dollars spent is not due to America's lack of concern about their excess body fat. On the contrary, Americans appear to be extremely concerned with body image as surveys indicate that up to 70 % of females and 40 % of males are on weight reducing diets. This obsession with weight control has lead to the lucrative industry of diet books and weight-control clinics making over $51 billion annually.

TABLE 7.1

Various Health Problems Related to Overfatness

✓ **Coronary Artery Disease**	✓ **Early Death**
✓ **High Blood Pressure**	✓ **Surgical Complications**
✓ **Type II Diabetes**	✓ **Kidney Problems**
✓ **Cancer (colon, rectum, prostate, breast, uterus, ovaries, and gall bladder)**	✓ **Cirrhosis of the Liver**
	✓ **High Blood Cholesterol**
✓ **Osteoarthritis**	✓ **Stroke**
✓ **Gout**	✓ **Varicose Veins**

Annual Health Care Cost of Overfat-Related Diseases in the United States

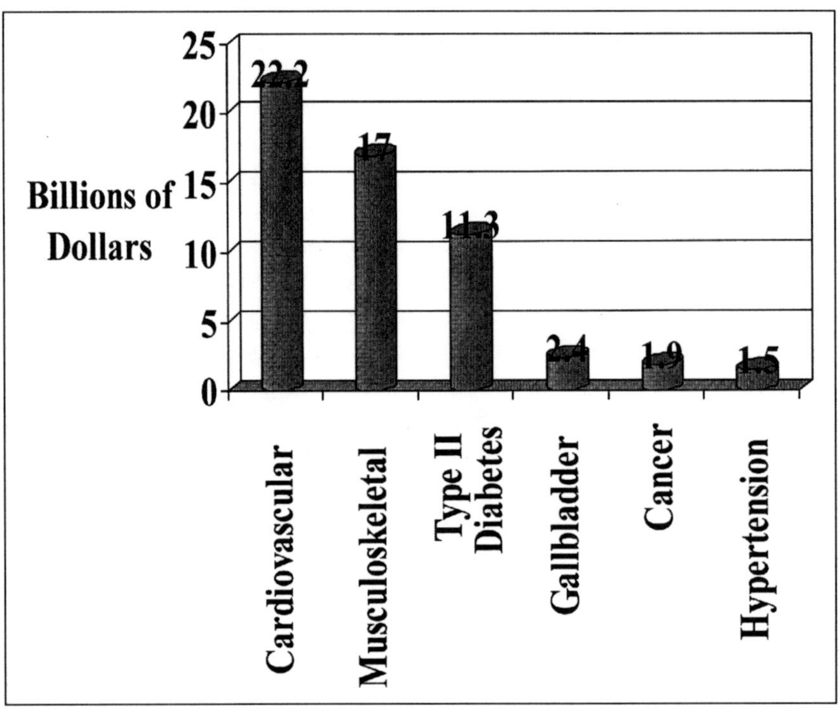

Body Composition

Human body mass can be categorized into two main groups. The first is lean body mass which consists of muscles, bones, and organs. The second body mass category is fat which is subdivided into essential and storage fat.

Essential Fat

A certain amount of body fat is required for normal physiological functioning. This fat is stored in the marrow of the bones and lipid tissues of the nervous system. In addition, important storage sites of fat are located in or around the heart, muscles, lungs, liver, spleen, intestines, and kidneys. Included in this essential fat is the sex-specific fat for females important for hormonal production, menstruation, bone development, and childbearing. Essential fat 3–5% for males and 10–15% for females.

TABLE 7.2

Body Fat Percentage Standards for Males and Females

STANDARDS	MALES	FEMALES
Unhealthy	**<5%**	**<15%**
Athletic	**5 - 10%**	**15 - 20%**
Health Fitness (Good)	**11 - 15%**	**21 - 25%**
Acceptable	**16 - 20%**	**26 - 30%**
Overfat	**>20%**	**>30%**

Storage Fat

Fat that accumulates in fat tissue (adipocyte) is termed storage fat. Some storage fat is located in the visceral, but it is mostly located under the skin and is called subcutaneous fat. Storage fat serves as a body insulator against the cold and as protective padding against physical trauma. In addition, it functions as an energy reserve. With just 4.5 kg (10 lbs.) of storage fat, a 68 kg (150 lbs.) male has enough stored energy (35,000 Calories) to run 296 miles nonstop at a nine minute per mile pace.

The acceptable percentage of body fat is different for males and females as well as for those attempting to gain health fitness or are competitive athletes (see Table 7.2). There are however, minimal and maximal standards of acceptable body fat to maintain a healthy body.

Assessment of Body Composition

The accurate appraisal of body composition is an important component of health fitness. There are two general techniques used to determined body composition. The first is direct evaluation through a process of chemical analysis and dissection. For obvious reasons, this method is only performed on animal carcasses or human cadavers.

The second technique utilized for evaluating body composition is through indirect analysis using a variety of methods. A quick review of the most common methods and their limitations are listed below:

1. **Hydrostatic Weighing.** Hydrostatic weighing, or underwater weighing, is designed around Archimedes' principle of density. Fat has a density of 0.90 grams per cubic centimeter as opposed to average lean tissue which has a density of 1.10 grams per cubic centimeter. This means that fat has a greater volume per weight allowing it to float in water where as lean mass with a higher density will sink. During hydrostatic weighing the individual is submerged under water where weight is measured. The greater amount of body fat, the less lower the underwater weight will be. Several calculations must be made to determine the percentage of body fat. This procedure is the most accurate to date; however, it requires a considerable amount of time, equipment, and skill by the technician.

2. **Skinfold Measurement.** The second best method for determining body fat percentage is the measurement of specific subcutaneous fatfold sites. This technique is based on the knowledge that in young adults 50 % of total body fat is located under the skin. As one age, the percentage of subcutaneous fat to total body fat slightly decreases. The most common anatomical locations for taking skinfold measurements are the triceps (back of the upper arm), pectoralis (chest), suprailiac (hip), umbilicus (abdomen), thigh, axilla (side of ribs), and supscapular (upper back). All measurements are taken on the right side of the body in a standing position. A specially designed caliper is used to measure the skinfold thickness in millimeters. With a trained technician, the results are usually within two percentage points of hydrostatic weighing.

3. **Bioelectrical impedance.** Bioelectrical impedance is based on the fact that electrical flow is facilitated by high electrolyte water content of the cells of the body. Since fat cells have low water and electrolyte content, impeded electrical flow will be directly related to the quantity of body fat. A painless, low voltage electrical current is introduced to the body and the resistance to the current flow is recorded and converted into a body fat per-centage. Though very simple to use, this method is not always reliable. Two factors that can greatly decrease the accuracy are the body hydration levels and room temperature.

4. **Body Mass Index.** Body mass index (BMI) is derived by dividing an individual's weight in kilograms by height in meters squared. This method is somewhat better than the old height/weight chart used by many insurance companies; however, it still does not consider the proportional composition of the body. The body mass index is easy to

calculate and shows if a person is at a high risk for developing cardiovascular diseases. The American College of Sports Medicine has stated that a BMI of 21 to 23 for women or 22 to 24 for men is considered desirable. When BMI exceeds 27.3 for women and 27.8 for men, there is a significant increased risk of cardiovascular disease.

Factors Affecting Body Fat

There are two factors affecting a person's amount of body fat (see Figure 7.2). The first factor is genetics which plays a role in predisposing a person to gaining excess weight and becoming overfat. However this in and of itself will not cause excess accumulation of body fat. This person must also be in a sedentary environment with access to food above the daily physiological requirements. Lifestyle is the second factor in affecting the development of overfatness and contributes to 75 % of the body fat a person will acquire. This factor includes living environment, social economical status, activity level, and accessibility to high caloric foods.

FIGURE 7.2

Factors That Contribute to Body Fat

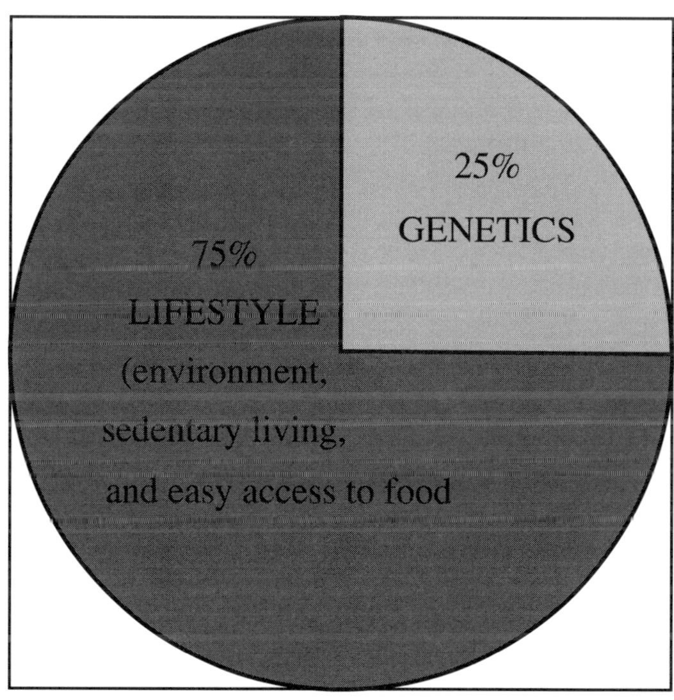

Components of Daily Caloric Expenditure

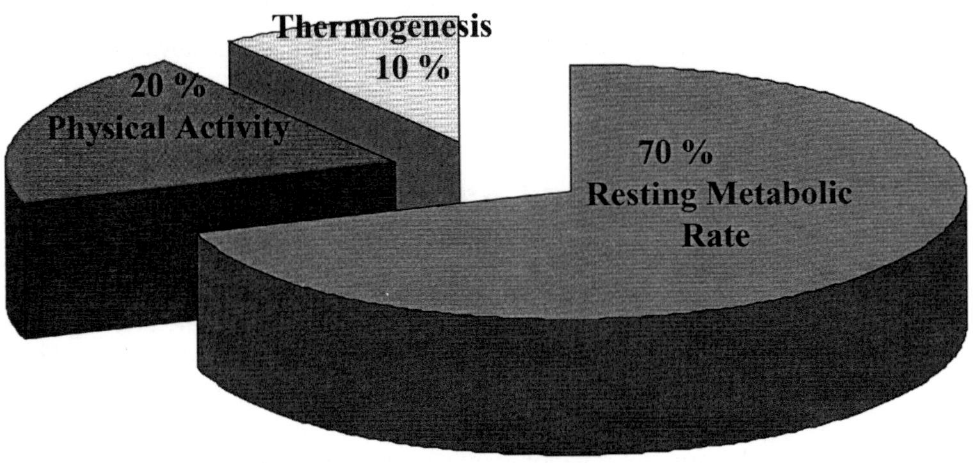

Daily caloric expenditure determines what effect lifestyle has on body fat accumulation. There are three components that determine the amount of calories burned each day (see Figure 7.3). The first component is thermogenesis which utilizes 10 % of daily caloric requirements. This energy is used by the body to break down consumed food into its parts and to transport them throughout the body. The second component in burning calories is physical activity which includes such things as exercise, gardening, and housework. On the job physical exertion also fits in this category making a combined total caloric expenditure of only 20 %.

The third component of daily caloric expenditure is resting metabolic rate and is composed of basal, sleeping, and resting (e.g. sitting, reading, and watching TV) metabolisms. This component burns 70 % of the calories expended each day with basal metabolic rate (BMR) using the most. BMR is the minimum level of energy required to maintain bodily functions in a resting, awake state and is directly related to the amount of lean body tissue. Women have a 5 – 10 % lower BMR than men of the same age and weight, due to their higher percentage of body fat. Also, as one age, BMR declines mainly because of a lower physical activity level leading to a decrease in lean body tissue. Clearly the best way to expend more calories each day is to increase lean body mass through regular exercise.

Energy Balance Equations

The three energy balance equations state that for body weight to be stable caloric intake must equal expenditures (see Figure 7.4). These equations only look at total body weight in regards to caloric consumption and expenditures; they do not specify the type of body mass (fat or muscle) lost or gained. For someone with a sedentary

FIGURE 7.4

The Energy Balance Equations for Body Weight

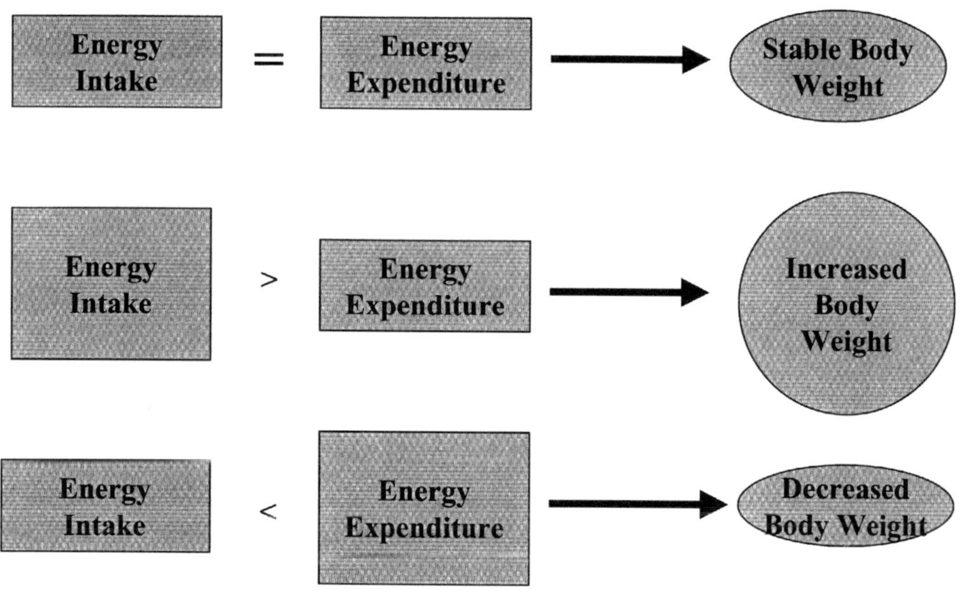

lifestyle, if caloric intake exceeds expenditure body weight increase will be in the form of storage fat. However, if caloric consumption is greater than output due to intense physical activity, then the added weight may be in the form of larger muscles and denser bones.

There are three ways in which an individual can achieve a decrease in body weight. The first method is to decrease caloric intake while maintaining the present energy expenditure level. This is the least desirable method of losing body weight because the majority of weight loss comes from water storage and lean tissue not just from adipocytes. When following a low calorie diet, the body perceives the condition as one of famine and tries to conserve body fat stores. To accomplish this, BMR is slowed allowing the body to survive with fewer daily calories. Once the goal weight has been reached, the amount of calories needed to maintain this weight would be less than before going on the diet. This causes the individual to either have to maintain this low caloric intake or gain back the weight with a greater amount of body fat than before starting the dieting. Research has found that up to 66 % of the weight lost from just dieting is regained within one year, and all is regained by five years. In addition, a low caloric diet will not be able to meet the Recommended Dietary Allowances for vitamins and minerals which will lead to deficiencies and possible disorders.

The second method to decrease body weight is by maintaining present caloric consumption and increasing energy expenditure. This technique keeps the body from going into the "starvation mode" and will create additional lean tissue causing an increase in BMR. With the caloric intake staying stable, one should never feel "deprived" of eating and resent the weight loss process. Increased caloric expenditure in the form of exercise also leads to improvements in other health fitness areas. Generally, maintaining caloric intake and increasing output is the best method for those needing to loose less than 4.5 kg (10 lbs.) of fat. In addition, this method is ideal for athletes where a proper diet of high nutritional value is required for optimal performance.

The final way to lose weight is to slightly decrease caloric intake (< 500 Calories below regular daily amounts) and increase the amount of calories burned each day (300-500 Calories) through exercise. This method works best for those needing to lose over 4.5 kg (10 lbs.) of fat weight. This combined approach is much less likely to cause hunger pains. Also, exercise protects against the loss of lean tissue and enhances the utilization of fat from the adipocytes.

Benefits of Exercise for Fat Reduction

Regular exercise with or without diet restriction presents many positive benefits for the individual needing to reduce body fat. The best type of exercise for fat loss is a combination of strength training and aerobic activity (see Figure 7.5). Strength training increases lean body mass causing an elevation in BMR. This leads to a higher daily caloric expenditure because lean tissue is more metabolically active

FIGURE 7.5

The Effects of Eight Weeks of Strength Training, Aerobic Exercise, and/or Low Caloric Diet on Body Composition in Obese Women

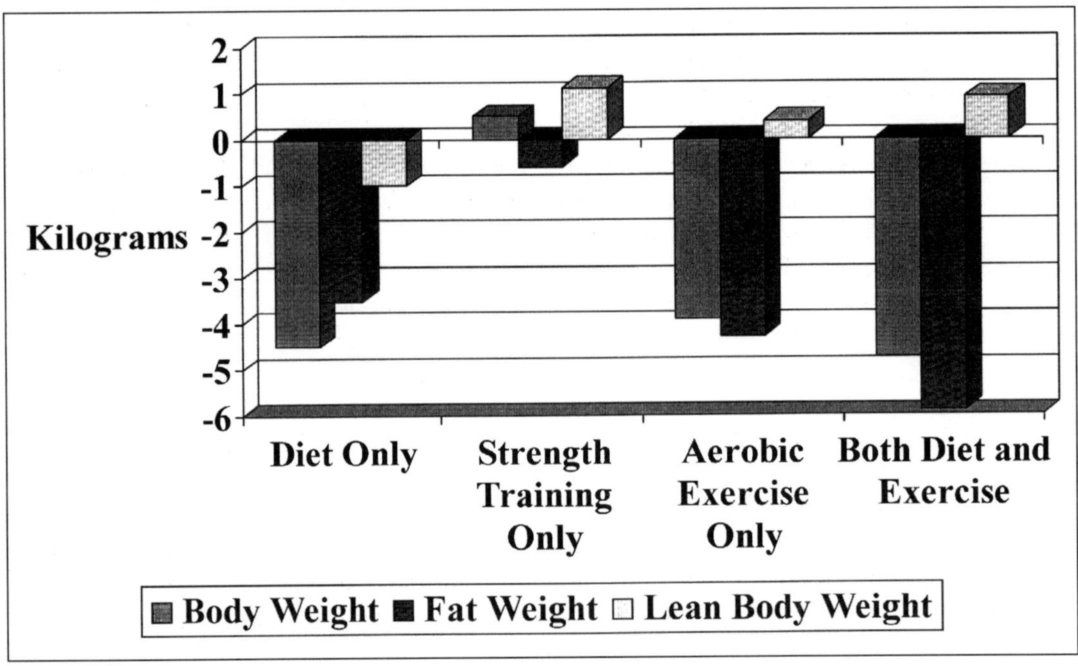

FIGURE 7.6

The Percentage of Fat and Carbohydrate Metabolized During Aerobic Activity

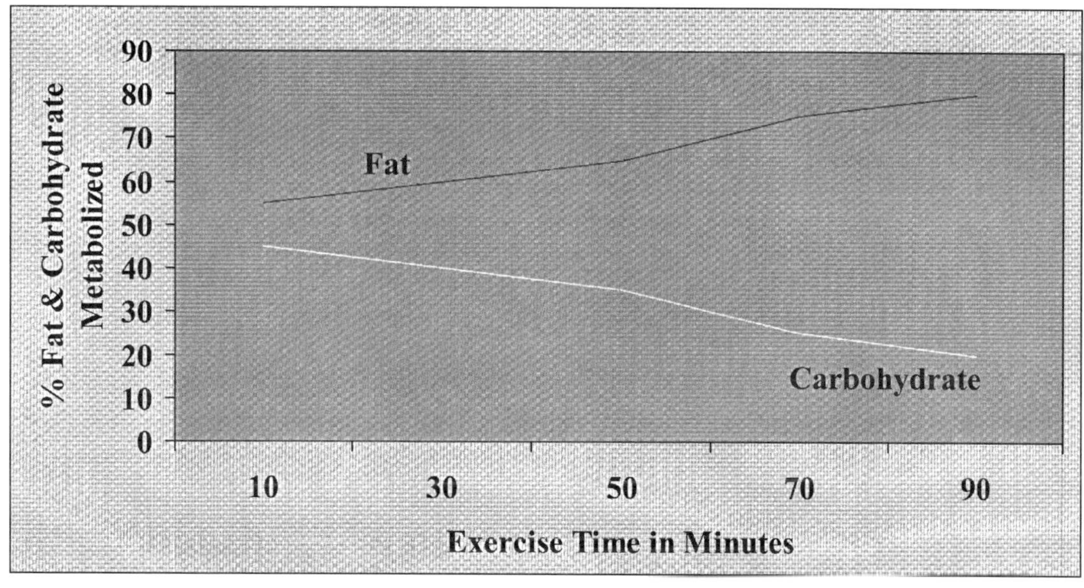

than fat tissue. On the other hand, aerobic exercise increases the mobilization and utilization of fat from fat stores (see Figure 7.6). Other benefits from aerobic activity include: an improved cardiovascular system, a decrease in blood pressure, and an increase in HDLs (Chapter 4). With regular exercise, it is possible to increase body weight but actually fit in smaller size clothing. This is possible because fat has a greater volume per weight than muscle. Weight is not a good indicator of fat loss; a better method would be by clothing fit.

Many individuals seek to reduce body fat at specific anatomical locations. This spot reduction is based on the belief that an increase in skeletal muscle activity under the fat storage site will facilitate this fat's utilization. Unfortunately this does not occur; the mobilization of fat to be used for energy comes first from the most metabolically active fat stores. Basically this means that the last place where fat was deposited will be the first fat used, and no type of exercise or special equipment will cause otherwise.

Eating Disorders

Only overfat individuals need to lose weight in the form of fat. Ideal body weight is based on body composition and the attainment of optimal health fitness rather than on body shape or weight. Many healthy–weight or thin women and men are striving to lose additional body weight to achieve some arbitrary weight. An individual's perception of body fatness and strong environmental pressures can cause some to believe they are overfat when in reality they are not. Poor self-concept and an unrealistic body weight goal (see Table 7.3) can lead to eating disorders and possibly death.

In the United States, the prevalence of eating disorders is between 5–10% of the population with approximately 90 % being female. The greatest number of people with eating disorders is found among young college females who comprise up to 20 % of this population. Three common eating disorders recognized by the American Psychiatric Association are 1) anorexia nervosa, 2) bulimia nervosa, and 3) binge eating disorder.

1. **Anorexia Nervosa.** Anorexia nervosa is an eating disorder in which a preoccupation with body weight leads to self-starvation. An anorexic individual possesses a great fear of becoming fat and has a distorted image of their body believing they are fat rather than actually underweight. These individuals typically begin a diet and at first feel in control and please with their weight loss. To increase the weight loss they combine exhaustive exercise, laxatives, and diuretics. As they lose excessive weight, their health begins to deteriorate. Typical problems from an anorexia behavior are amenorrhea, digestive problems, anemia, growth of fine body hair, depression, osteoporosis, and abnormalities of the immune system. Anorexics cannot overcome this disease by themselves; a professional trained in eating disorders is required. If help is obtained quickly, many of the physical complications can be reversed.

2. **Bulimia Nervosa.** Another eating disorder is bulimia nervosa and is more prevalent than anorexia nervosa. Bulimia is characterized by eating regular or large meals (binges) followed by intentionally

TABLE 7.3

Warning Signs of Eating Disorders

✗ A preoccupation with food, calories, and weight.
✗ Concern about being or feeling fat.
✗ Self-criticism of one's body.
✗ Secretly eating or stealing food.
✗ Unwillingness to eat in front of others.
✗ Periods of severe caloric restriction.
✗ Use of laxatives.
✗ Mood swings.
✗ An apparent preoccupation with the eating behavior of others.
✗ Wide fluctuations in weight over short period of time.
✗ Vomitus, or odor of vomitus in the bathroom.
✗ Consumption of large meals followed by trips to the bathroom.

emptying (purging) the food out of the stomach by self-induced vomiting, use of laxatives, or intense exercise. Bulimia occurs primarily in young women with a morbid fear of becoming fat. Most bulimics are healthy-looking, good students or athletes, extremely sociable, and pleasant. Medical problems associated with this disease include amenorrhea, tooth erosion, cardiac arrhythmias, ulcers, and kidney damage. Bulimics need assistance in overcoming this disorder and require professional help from a trained eating disorder specialist.

3. **Binge Eating Disorder.** Individuals with binge eating disorder consume large amounts of food in a short period of time; however, unlike bulimia nervosa purging is not included. Binge episodes are often triggered by emotional and/or psychological events (e.g. loneliness and anxiety) instead of physical hunger. These binges usually occur in private and are followed by feelings of guilt, shame, and depression. This disease requires professional help from a trained specialist.

Summary

Overfatness is a serious situation in the United States causing a variety of health problems and early death. Health care costs for overfat-related illnesses are over $56 billion annually. The human body requires a minimal amount of fat for proper physiological functioning. Excess fat is stored in many locations with approximately 50 % in the subcutaneous area under the skin. Lifestyle plays the largest role in affecting body fat accumulation. However, by following the energy balance equations, body mass can be modified. The best method for losing body fat is by a slight reduction in caloric intake and adding exercise consisting of both strength training and aerobic activity. Eating disorders are prevalent on college campuses especially among female students and athletes.

Questions

1. Distinguish between body weight and body composition.

2. What are acceptable body fat percentages for males and females?

3. List and explain three techniques for determining body fat percentage.

4. Define anorexia nervosa, bulimia nervosa, and binge eating disorder.

5. List five health problems related to excessive body fat.

6. What are the three methods for losing body weight?

7. Explain why lean tissue has an effect on basal metabolic rate.

8. List the three ways in which energy is expended in the body.

8

The Physically Active Lifestyle and Aging

Learning Objectives

This chapter recommends establishing a regular exercise routine and sticking to it not for just a few days, weeks, or months but for a lifetime. The message is that regular exercise should be developed as one would develop a lifelong health habit. Guidelines for regular exercise and recommendations for an active lifestyle are encouraged. When the chapter is completed you should know:

1. Guidelines for disciplining yourself to have a consistent exercise program.

2. Daily physical activities that can be added to improve health fitness.

3. The benefits of being physically active as one ages.

4. The benefits of being physically active in reducing health problems.

Key Terms

Cognitive Functioning

Physically Active Lifestyle

Socio-Cultural

World Health Organization

Making Exercise a Regular Habit

Regular physical exercise, performed on most days of the week, reduces the risk of dying prematurely. In addition, it promotes psychological well being and decreases the risk of cardiovascular diseases that claim the lives of one million Americans every year. Interestingly, even with the knowledge of these benefits, many Americans do not exercise on a regular basis; less than 15 % of adults exercise three or more times a week. The greatest benefits from exercising regularly occur later in life; thus causes many young adults not to realize its importance until they are faced with a heart attack or stroke.

Starting an exercise habit is not difficult, however maintaining it is. The following are guidelines that can assist in making exercise a regular habit.

1. **Place into the weekly schedule the days and times designated for exercise.** Determine the days of the week that will best accommodate an exercise program, and designate them as "exercise days". Also, find a suitable time during each of these exercise days to reserve for the workout. There are only minimal differences in benefits from exercising during the morning, noon, or night; the key is choosing a time that is most convenient and that causes the least disruption to the daily schedule.

2. **Discipline one's mind to plan on exercising during the schedule time regardless of feelings.** Only an illness or important family event should cause a missed workout. When sessions must be missed more than once during a week, they should be rescheduled. If there are frequent interferences with a particular time, the schedule should be adjusted to avoid the conflict. Missing an exercise session as little as once a month is not a concerned, but missing a workout once a week could soon become missing twice a week and eventually lead to the loss of the exercise habit.

3. **Modify exercise rather than discontinue it when an injury occurs.** If an injury prevents the performance of the normal exercise routine then the workout should be modified, not eliminated. There are many types of aerobic and strength training activities that can be utilized without aggravating or making the injury worse. If the disability is permanent, then appropriate modifications should be made to the exercise program. An injury must not be allowed to stop the acquisition of health fitness.

4. **Remember that the more hectic and busier the daily schedule is, the more important the exercise session becomes.** Exercise reduces stress and enhances mental concentration allowing a greater amount of work to be preformed more efficiently. Solutions to problems come faster, and less mental fatigue occurs.

5. **Continue to exercise regularly into old age.** The need for exercise increases with age because low activity causes muscles to atrophy, bones to weaken, and body fat to increase. A continued sedentary lifestyle often leads to disability causing one to no longer be self-sufficient or have the freedom to move about.

6. **Wear appropriate exercise attire including footwear and undergarments.** Wearing correct clothing for the type of activity will enhance the enjoyment of the exercise and decrease the risk of injury. Also, use the exercise attire just for working out to increase its longevity. Running shoes should be used strictly for running and not for everyday activities. These shoes need to be replaced after every 500 miles of use due to the loss in their protective cushioning. Once the shoes are replaced, they may be used for normal footwear.

Developing A Physically Active Lifestyle

Following an exercise program on a regular basis is only part of having a physically active lifestyle. Exercising one hour per day for six days per week does not mean the other 162 hours in the week can be inactive. While six hours of exercise per week will markedly enhance health fitness, modification of other daily activities will further improve one's health especially if overfatness is a problem. The following are ways to create a more physically active lifestyle:

1. **Use the stairs rather than the elevator or escalator.** When a choice can be made at work, school, or shopping, use the stairs. By just climbing two flights of stairs per day, enough calories will be expended to burn approximately .5 kg (one pound) of fat in a year.

2. **Park farther away from the store or office to increase the walk to and from the car.** If riding the bus or train for commuting, get off one stop earlier for an extended walk. Walking two additional blocks per day allows an individual to expend enough calories in a year to lose over 2.5 kg (5 lbs.) of fat.

3. **Walk or bicycle; don't drive/ride.** Try to make walking or bicycling the first choice for transportation. Not only will health fitness improve, but also money will be saved from less gasoline used, and the environment will be cleaner.

4. **Avoid laborsaving devices.** Push the lawn mower to cut the grass instead of using a riding or self-propelled mower. When playing golf walk and pull or carry the golf clubs. By playing 18 holes of golf without riding in a cart, up to five miles could be walked.

5. **Develop hobbies that require physical effort**. Choose leisure activities or hobbies that involve frequent physical movements such as: gardening, dancing, and hiking. Also participate in leisure sports such as: bowling, horseshoes, sailing, and table tennis. While these activities will not produce fitness gains alone, they will increase the amount of physical effort used during the week.

Physical Activity and Aging

At the beginning of the 20th century only four percent of the United States population was over 65 years of age. By the year 2020 this group of senior citizens will increase to 20% of the American population. This increase in the number of older adults may have dire consequences for society if they do not embrace physically active lifestyles. Advancing age is associated with potential sensory, motor, and cognitive changes impacting elderly people's ability to function independently in society.

In recent years, researchers have established that hereditary factors play an important role in how one ages. However, additional factors influence the aging process and can modify the outcome. The largest aging modifier is a healthy lifestyle consisting of regular physical activity and good nutrition. By participating in regular physical activity throughout one's life a number of physiological, psychological, and socio-cultural benefits occur.

Cardiovascular Fitness. Maximal oxygen utilization was thought to decline at about 10 % per decade with advancing age. However, recent studies have shown that by maintaining one's aerobic activity level, little decline occurs in aerobic capacity. In addition, when sedentary senior citizens increase, even modestly, their levels of physical activity, drastic improvements in cardiovascular function will result.

Blood Pressure. Hypertension is a serious medical problem that afflicts more than 20 million older Americans. Both systolic and diastolic blood pressure increases with advancing age. However, with a low intensity walking program, significant decreases in both systolic and diastolic blood pressure have been found in hypertensive adults over 60 years of age.

Blood Lipids. Aging is associated with an increase in cholesterol levels that can cause development of cardiovascular disease. Studies have found that older individuals on regular aerobic exercise programs have favorable cholesterol levels as compared to sedentary individuals of the same age. An additional area that effects blood lipid levels is body composition. Individuals with healthy body fat percentages have a more positive cholesterol level as compared to overfat persons of the same age.

Muscular Endurance and Strength. Muscle endurance and strength decline with advancing age. Adequate levels of muscular strength are critical for the performance of daily living activities. Until recently strength training was considered potentially dangerous for the elderly. However, the latest studies have shown that strength training can be performed safely in both males and females as old as 90 years of age with strength gains over 100 %.

Flexibility. Aging is associated with changes in the elasticity of connective tissue resulting in decreases in range of motion of the joints. Most declines in flexibility are directly linked to a decrease in physical activity. Stretching exercises in the older individual will increase flexibility and joint range of motion.

Balance. Age related declines in postural stability and dynamic balance are risk factors for falls and injuries in the older adult population. An exercise program emphasizing walking, flexibility training, and strength training will improve balance and decrease body swaying in senior citizens thus increasing their mobility and self-sufficiency.

Depression and Anxiety. The incidence of depression increases significantly with age and is associated with a decline in physical activity. Participation in regular physical exercise has been shown to reduce mild to moderate levels of depression and to beneficially effect Mood State and anxiety.

Cognitive Functioning. Decrements in cognitive processes occur as one ages. However, there is a wide variation in decreases depending upon the cognitive task. Regular exercise has been shown to postpone age-related declines in Central Nervous System processing speed and in both fine and gross motor performance. Elderly people, who participate in sporting activities that require adjusting to a constantly changing environment (racquetball and tennis), are able to maintain higher levels of coordination and better reaction times than sedentary individuals of similar age.

Social Functioning. Regular physical activity enhances many socio-cultural variables in the elderly. In 1997, the World Health Organization published a summation of their findings (see Table 8.1) indicating that the social benefits of exercise for older individuals are significant.

TABLE 8.1

A Summary of the Social Benefits of Physical Activity for Older Persons by the World Health Organization, 1997

✳ **Empowering Older Individuals in Society.**

✳ **Enhancing Social and Cultural Integration.**

✳ **Allowing the Formation of New Friendships.**

✳ **Widening Social and Cultural Networks.**

✳ **Permitting Role Maintenance and New Role Acquisition.**

✳ **Encouraging Intergenerational Activity.**

Summary

Having a physically active lifestyle requires more than just maintaining a regular exercise program. It also includes making all aspects of one's life more physical. Today there are many modern devices for saving time and labor; however, their use is not recommended so that one may reap the benefits of being more physically active. During the aging process many bodily changes occur, but with regular exercise most of the decline is slowed. Even though an individual has been sedentary into later years, increasing physical activity will reverse many harmful physical and psychological changes and will enhance social functioning.

Questions

1. Discuss the recommendations for increasing activity beyond a regular exercise program.

2. List three ways to make exercise a regular habit.

3. What will be the percentage of Americans who are 65 or older in the year 2020?

4. Name three physiological benefits of continuing physical activities into old age.

5. List the social benefits of regular physical activity for the elderly.

9

Basic Nutrition for Health Fitness

<table>
<tr><td>

Learning Objectives

This chapter discusses basic nutritional needs for health fitness. The need for and function of carbohydrates, fats, protein, vitamins, minerals, and fluids in the body are explained. Simple and complex carbohydrates, fiber in the diet, and triglycerides are detailed in terms of their affects on the health of the body. Basic food groups, recommended daily menus, calories, and nutritional information are provided in chart and table form. When you complete this chapter you should:

1. Understand the role and function of the seven basic nutrients in the body.

2. Know the food sources and recommended daily intake of the seven nutrients.

3. Know the seven basic food categories.

4. Know the health concerns and benefits of proper nutrition, especially as they relate to carbohydrates, fats, and protein.

</td><td>

Key Terms

Calories

Carbohydrate

Complex Carbohydrate

Fiber

Insulin

Minerals

Protein

Saturated Fat

Triglycerides

Unsaturated Fat

Vitamins

</td></tr>
</table>

Eating For Health

Health is very dependent upon the food and fluids consumed. Poor eating habits have been associated with hyperactivity in children, physical fatigue, mental fatigue, depression, stunted growth, and numerous other problems. Five of the 10 leading causes of death have been associated with diet: 1) heart disease, 2) cancer, 3) atherosclerosis, 4) stroke, and 5) Type II diabetes.

The body is composed of the elements outlined in Table 9.1. These elements are constantly being used to provide energy for the body, to build new cells, to repair old cells, and to regulate the various body processes. Since the body is unable to absorb these elements through the skin, they must be gained through daily consumption of foods and fluids.

Seven Basic Nutrients

Nutrients are chemical parts of food and liquids that have specific functions in the body. The two overall functions of nutrients are 1) to provide essential elements the body needs to sustain life and 2) to provide energy for the body. The essential elements needed for health fitness are found in seven basic nutrients (see Table 9.2).

TABLE 9.1

Percentage of the Basic Elements of the Body

NON-METALLIC ELEMENTS		METALLIC ELEMENTS	
Element	Percent	Element	Percent
oxygen	65	calcium	1
carbon	18	phosphorus	1
hydrogen	10	sodium	1
nitrogen	3	potassium, chlorine, sulfur, magnesium, and15+ others	1
TOTAL	96	TOTAL	4

The Seven Basic Nutrients

• **CARBOHYDRATES**

• **FIBER**

• **FATS**

• **PROTEINS**

• **VITAMINS**

• **MINERALS**

• **WATER**

1. Carbohydrates

The primary function of carbohydrates is to provide all the cells of the body with energy. While most cells use a combination of both fats and carbohydrates for energy, the central nervous system (i.e. brain and nerve tissue) uses only carbohydrates for energy. Muscle cells use a combination of carbohydrates and fats for energy. However, in high energy requiring activities such as in sprinting and strength training, carbohydrates are used exclusively for energy since fat metabolism is a slower energy producing process, and it requires oxygen.

Carbohydrates must be present in the diet for fats to be completely utilized for energy. If a diet is low in carbohydrates, then proteins from the skeletal muscles will be broken down and converted into carbohydrates to meet this need. This procedure of catabolizing muscle tissue will cause problems in the body. On the other hand, excess carbohydrates in the diet are converted to fat and stored.

Types of Carbohydrates.   There are two basic types of carbohydrates: 1) simple, which are mainly mono-, and disaccharides and 2) complex or polysaccharides. Simple carbohydrates consist of sugars, (see Table 9.3) and complex carbohydrates consist of starch and cellulose (fiber). In processed foods, simple carbohydrates go under the names of brown sugar, honey, corn syrup, levulose, and dextrose. Regardless of the name, the body handles them all as sugars and must convert each of them into the usable form called glucose. The current American diet is too low in total carbohydrates but too high in simple sugar. Each year Americans have been consuming more processed sugar.

TABLE 9.3

The Six Common Simple Carbohydrate Types

TYPE	SOURCE
❑ Sucrose (table sugar)	Sugar beets and sugar cane
❑ Fructose	Fruits and honey
❑ Galactose	Milk
❑ Lactose	Milk
❑ Maltose	Grains and cereals
❑ Glucose	Fruits, corn syrup, and honey

A complex carbohydrate is composed of many sugars (50 to over 1000) combined together to form a complex molecule. Complex carbohydrates are primarily found in vegetables, grains, seeds, and fruits. Of the carbohydrates in the American diet, approximately half come from starches and half from sugars. Experts recommend an increase in total carbohydrate intake from 45 % to 60 % of our diet. The amount of carbohydrates received from starch should be increased from 50 % to 75 %.

Complex vs. Simple.   Complex carbohydrates in the form of starch are not directly absorbed into the blood; only simple sugars are. When complex carbohydrates are eaten, enzymes in the mouth, stomach, and small intestine begin to break the carbohydrates down into their simplest form (i.e. sugar). After this process has been accomplished in the small intestine, the sugar can pass through the intestinal wall and is absorbed into the blood stream where the liver converts these sugars into glucose which is the form body cells can use. To break a starch down into its simplest form takes several hours. Since starches take a long time to be digested, they pass slowly into the blood stream allowing blood glucose levels to fluctuate gradually and minimally.

On the other hand, when a simple carbohydrate is consumed it passes very rapidly (less than 60 minutes) through the digestive tract to the blood stream and causes a sudden increase in blood glucose levels. This rapid increase stimulates a large amount of insulin to be released from the pancreas. Insulin's function is to transport the glucose out of the blood and into the cells of the body. If the cells do not require glucose for energy, the glucose is changed into fat and stored in adipocytes. In this way, high-sugar diets can lead to increased body fat.

The excess insulin released due to the rapid increase of blood glucose level causes the liver to produce triglycerides and very low-density lipoproteins and to release them into the blood. This can accelerate atherosclerosis (Chapter 4) making excess dietary sugar a role player in cardiovascular disease. In addition, the excess insulin will transport too much glucose out of the blood and into the cells causing low glucose levels (hypoglycemia) leading to weakness and hunger. The person then eats again, when it is really not necessary, demonstrating how a high-sugar diet leads to a craving for more sugar.

Over a period of years, the up-and-down effect of high blood glucose and high insulin levels followed by low blood glucose and low insulin can cause the cell to become resistive to insulin. Also, the insulin generating cells of the pancreas can become over-worked and slow their production of insulin. In time the blood glucose level remains elevated and a condition of diabetes develops. Most Type II diabetes is due to this factor and is caused in part by the person's dietary habits, overfatness, and lack of exercise.

When the diet contains excessive amounts of refined sugar that is added to soft drinks, bakery goods, and similar products, the body is being robbed of important nutrients. Table sugar is called "empty calories" because it contains no vitamins, minerals, or fiber. Also refined sugar sticks to the teeth and, with the bacteria present in the mouth, causes tooth decay (dental caries).

2. Fiber

Fiber is a type of complex carbohydrate of which cellulose is the most common form and is found primarily in fruits, vegetables, and grains. Since the body does not possess the enzymes necessary to break down fiber, it has no caloric value, and it passes through the small and large intestine relatively unchanged.

Function of Fiber.   Although several specific purposes of fiber have been established, much is yet to be learned about its importance in the diet. Fiber promotes regular bowel movements by absorbing water in the large intestine making the stools soft and large. Thus, elimination from the body is easily preformed with no straining or pushing. With a high fiber diet, constipation is rare and so are hemorrhoids.

High-fiber diets also quicken the transit time of food from intake to elimination. Waste products of metabolism stay in the large intestine three times longer with a low-fiber diet. This allows the waste products, some of which may consist of toxic chemicals and carcinogenic agents (i.e. cancer causing), a longer time to irritate the large intestine. This may be a strong factor in diverticulosis, colitis, and colon cancer. Authorities recommend one to two bowel movements a day.

Lack of fiber in the diet has also been associated with high blood cholesterol, appendicitis, gallbladder diseases, diabetes, and obesity. Fiber is absolutely an essential part of a healthful diet, not so much for what it does but for what it prevents.

Sources of Fiber.   The amount of fiber needed daily in the diet is 1 gram per 100 calories. A typical American eats four to eight grams daily, but vegetarians eat as much as 30 to 40 grams. Most current authorities recommend a safe amount of fiber of at least 25 grams per day. If a person does not normally have one to two bowel movements per day with soft stools, chances are there is not enough fiber in the diet. Like any nutrient, too much of a good thing can lead to problems. Excess fiber in the diet can lead to diarrhea, gas, and cause dehydration and mineral loss.

The sources of fiber should vary (see Table 9.4). Bran found in grain serves as one of the best sources of fiber. However, if the grain has been refined, the bran is lost and the fiber is gone. After a food is refined, some of the vitamins and minerals are restored (enriched) and occasionally additional vitamins are added (fortified), but the fiber is still absent. Therefore, unrefined grains are best in order to receive the fiber they possess. While bran from grains is a good fiber source, at least half the fiber should come from vegetables and fruits. Cooking vegetables and fruits causes them to lose some of their fiber. Therefore, we should eat vegetables and fruits raw as much as is possible.

Breakfast cereals are an excellent source of fiber if properly selected. Table 9.5 lists the breakdown of carbohydrates in selected high-fiber cereals. A breakfast of Corn Bran, Multi Bran Chex, or 40 % Bran Flakes topped with strawberries or a banana can start the day off with low sugar, high starch, and high fiber.

TABLE 9.4

Fiber Content in Selected Fruits, Vegetables, and Grains

FOOD	SERVING	FIBER GRAMS
√ Broccoli	1 medium stalk	6
√ Apple	1 medium	5
√ Almonds	1/4 cup	5
√ Cabbage	1/2 cup	4
√ Peas	1/2 cup	3
√ Potato	1 medium	3
√ 100 % whole wheat bread	2 slices	2
√ Banana	1 medium	2
√ Orange	1 medium	2
√ Strawberries	1/2 cup	1
√ Carrots	3/4 cup	1

TABLE 9.5

Fiber Amounts in Selected Breakfast Cereals

	CEREAL (1 Serving)	TOTAL CHO (grams)	STARCH (grams)	SUGAR (grams)	FIBER (grams)
TM	**All Bran**	21	7	5	9
TM	**Raisin Bran**	31	11	14	6
TM	**40% Bran Flakes**	23	13	5	5
TM	**Multi Bran Chex**	25	15	6	4
TM	**Cracklin Bran**	21	10	7	4
TM	**Shredded Wheat**	19	16	0	3
TM	**Grape Nuts**	23	17	3	3
TM	**Cheerios**	20	19	1	2
TM	**Total**	23	18	5	2
TM	**Special K**	20	16	3	1
TM	**Fruit Loops**	25	12	13	0
TM	**Apple Jacks**	26	12	14	0

3. Fats

Fats provide most cells of the body with energy. At rest about 66 % of the cells' energy comes from fat and 34 % from carbohydrates. Fats are also important as carriers of vitamins (A, D, E, and K), as protectors of vital body organs, and as insulators of the body. They are also an important part of cell walls and hormones, and serve to depress appetite. Excess fat in the diet is stored in adipocytes.

Some fat in the diet is essential for normal body functioning; however, most American consumes too much fat everyday. At the turn of the century, fat accounted for 25 to 30 % of Americans' daily caloric intake compared to 40 to 45 % today. Teenagers tend to have poor dietary habits, and often 50 % of their diets are fat. Problems from excess fat intake include overfatness and elevated blood fats leading to cardiovascular disease (Chapter 4).

Types of Fats.   Triglycerides are the most common fats in the diet and are either saturated or unsaturated. Saturated triglycerides (saturated fats) are found primarily in animal products such as beef, pork, lamb, lobster, shrimp, milk, cheese, butter, cream, and eggs. Two plant sources that also contain high saturated fat are coconut and palm oil. Although fats are essential for health, saturated fats are not and should be reduced in the diet.

Unsaturated fats, which are polyunsaturated and monounsaturated, do not appear to cause health problems to the same degree as saturated fats. They are found primarily in vegetables but also are in oils (e.g. safflower, canola, corn, peanut, olive, soybean, and cottonseed) and margarine used in baking and cooking.

Another type of fat is cholesterol that is required by the body and is manufactured by the liver. Cholesterol is found in most cells and is a basic structure for several hormones. It is normally present in the blood attached to lipoproteins such as HDL and LDL.

Dietary Fat Sources.   The daily requirement for fat appears to be between 10 and 30 % of caloric intake. To evaluate fat intake, count the grams of fat eaten each day. Table 9.6 presents a recommendation for daily fat intake on the basis of one's situation. You should reduce your overall intake of fat by increasing low-fat foods and decreasing or avoiding fatty foods. By substituting low-fat meat and milk products, a person can significantly reduce fat intake. One baked potato is 100 calories, but one cup of French fries is 480 calories. One cup of whole milk is 150 calories compared to 85 calories in one cup of skim milk. The difference in calories is exclusively due to the fat in the fried potatoes and in the whole milk.

4. Protein

The major functions of proteins are to build, repair, and maintain cells. They form the major part of muscle tissue and are essential for muscle contraction. They are an important part of hemoglobin as well as numerous hormones. Excess proteins in

TABLE 9.6

Daily Amount of Fat Grams Intake According to Different Situations

THE PERSON IS/HAS:	Grams of fat/day
Overfat, sedentary, or has high blood lipids.	< 40
A history of heart disease in the family.	< 40
Acceptable body fat, average blood lipids, and regular exerciser.	40 - 60
Lean body fat, low blood lipids, and exercises vigorously daily.	60 - 80

the diet are converted to fat and stored. Normally the body does not use proteins for energy, but in emergencies, proteins can be converted to carbohydrates and used for energy.

The basic units, or building blocks, of proteins are amino acids. There are 23 different amino acids that can be combined in numerous ways to make a single protein. Ten amino acids cannot be synthesized by the body and are called "essential" amino acids. A complete protein is one that contains all the essential amino acids, whereas an incomplete protein does not contain all the essential amino acids. Sources of complete protein include eggs, milk, meat, fish, and poultry.

The average American consumes too much meat. Most people plan their meals around the meat they are serving and at restaurants, the main entree is usually meat. The daily need of protein is about 0.8 to 1.2 gram per kilogram of body weight for the normal adult and 1.5 to 2 grams per kilogram of body weight for the adolescent, the body builder, and the pregnant woman or nursing mother. This means that adults need only three to four ounces of protein per day to meet all their protein needs.

High-Protein Diet.   Several potential health problems are associated with a high-protein diet. Persons on a high-protein diet eat an excess amount of meat. The high-meat diet will result not only in excess protein consumption but also in high fat intake, with the fat being in the form of cholesterol and saturated fat. Beef and pork are 20 to 40 % fat, where as chicken and turkey without the skins and fish contain only 10 to 15 % fat.

The excess protein in the diet does not build excess cells; it is instead converted in the liver to fat and stored in adipocytes. When the body converts protein to fat, ammonia is produced. Since ammonia is toxic to the body, it is changed into urea and eliminated by the kidneys in the urine. Since urea is also toxic, water is drawn from the body cells to dilute it, and excess water is lost from the body in the urine. Therefore, a high-protein diet can cause the body to lose excess amounts of body water and leading to dehydration. Many persons on high protein diets enjoy quick and large weight losses. However, the weight loss is only water, not fat, and will return as soon as the diet is ended.

Vegetarian.   Many people, including current nutritional scientists, believe humans function best by only consuming vegetables, fruits, and grains. It is possible to receive all the essential amino acids without eating meat. A very important point, however, is that the vegetarian must emphasize a variety of foods in the diet, especially soybeans, peas, lentils, and beans, and must supplement these with cheese, milk, and eggs. When a variety of nonmeat foods are eaten, all the necessary protein can be attained. The only nutrient missing in a vegetarian diet is Vitamin B12 (from meat sources) which can be obtained through a vitamin supplement.

5. Vitamins

Vitamins are organic substances required for optimal health. They help regulate nearly all metabolic reactions, help convert carbohydrates and fat into energy, and assist with bone and tissue repairs. The body cannot manufacture vitamins so they must be ingested daily into the body. The Recommended Dietary Allowances (RDA) for vitamins has been established by the National Research Council of the Food and Drug Administration and is revised periodically as new research becomes available (see Table 9.7)

TABLE 9.7

RDA Amounts for Vitamins, Their Function and Major Source

VITAMIN	ADULT RDA	PRIMARY BODY FUNCTION	MAJOR SOURCE
A	5000 IU[1]	Tissue growth and repair, especially skin and all membranes; night vision	Vegetables, especially carrots, sweet potatoes, spinach; cantaloupe, watermelon
D	500 IU	Bone calcification	Milk, fish, sunshine since it is naturally formed in the body in reaction to the sun's rays
E	30 IU	An antioxidant to prevent cell membrane damage, may help to prevent scar tissue, may help to delay aging	Wheat germ, walnuts, sunflower seeds, almonds
K	1000 IU	Assists in the clotting of blood	Vegetables, especially spinach, cauliflower, cabbage
B-1 Thiamin	1.4 mg[2]	Aids in carbohydrate metabolism	Wheat germ, rice, peanuts, sunflower seeds, soybeans
B-2 Riboflavin	1.6 mg	Health skin, aids in metabolism to give us energy	Wheat germ, soybeans, cheese,
B-3 Niacin	18 mg	Converts fats and proteins to energy when needed, may help lower cholesterol	Soybeans, fish, chicken, sunflower seeds, wheat germ, sesame seeds
B-5 antothenic Acid	10 mg	Aids in nerve impulse transmission, aids in obtaining energy from foods	Soybeans, wheat germ, rice, broccoli, sweet potatoes, milk, fish
B-6 Pyridoxine	2.0 mg	Metabolism and synthesis of proteins	Wheat germ, sunflower seeds, soybeans, beans, rice, fish
B-12	.003 mg	Needed for the formation of red blood cells	Fish, dairy products, eggs, meat, (not found in plants)
C Ascorbic Acid	60 mg	Holds body cells together, strengthens blood vessels, helps wounds to heal, may fight infections and viruses	Fruits, especially citrus, melons and strawberries; vegetables, especially broccoli, spinach, greens, cauliflower
Biotin	.05 mg	Aids in fatty-acid production and metabolism of proteins	Soybeans, wheat germ, peanuts, mushrooms, fish, eggs
Folacin	.4 mg	Production of essential body proteins such as hemoglobin, RNA, and DNA	Beans, spinach, soybeans, peanuts, wheat germ, oranges

[1] International Unit
[2] Milligrams

Although taking less than the RDA for an extended period of time may lead to a health problem, an excess of the RDA will not improve one's health. In fact, consistent excess of certain vitamins can lead to medical problems. Americans spend nearly $500 million each year on vitamin and mineral supplements most of which exceed the body's needs. In the United States, health problems due to a vitamin deficiency are rare.

Types of Vitamins.   Two major categories of vitamins are fat-soluble and water-soluble. Fat-soluble vitamins are A, D, E, and K and must be ingested with fat. If fat is not present, they will pass out of the digestive tract and will not be absorbed into the body. Therefore, when taking a vitamin supplement, it should be taken with a meal. Once absorbed into the body, fat-soluble vitamins can be stored in fat, muscles, and the liver to be used when needed.

Water-soluble vitamins are readily absorbed into the body but are just as readily passed out of the body in urine and perspiration. They are not stored in the body and, therefore, need to be ingested daily. Water-soluble vitamins include B, B2, B6, B12, C, niacin, pantothenic acid, biotin, and folacin.

Antioxidants.   During the process of aerobic metabolism free radicals are created. A free radical is a chemically unstable molecule or ion containing an unpaired electron (radical) and can exist independently (free). Free radicals will react with other molecules (fats, protein, and DNA) to replace its missing electron. This process causes damage to cell membranes and mutation of genes. Free radicals have been linked to aging, cancer, cardiovascular disease and arthritis.

Antioxidants protect the body from free radical damage by limiting their formation. In addition, antioxidants can donate electrons to free radicals as well as repair damage caused by free radicals. The most noted antioxidants are vitamin E, vitamin C, and beta-carotene, which are found in vegetables and fruits. Regular exercise improves the body's normal physiological antioxidant defense system. However, repeated strenuous exercise may cause an increase in free radical formation and cellular damage.

6. Minerals

Minerals comprise four percent of body weight and are essential for many vital body processes. Not only are minerals needed for health, but also the proper balance among minerals is critical. Minerals work closely together, and each affects the other. Mineral imbalances can cause dehydration, heart attacks, and many other health problems.

Minerals are divided into two categories: 1) major minerals and 2) trace minerals. Major minerals are those for which the body has a large daily requirement of over 100 milligrams and include calcium, phosphorus, sodium, chlorine, potassium, magnesium, and sulfur. The specific role of the major minerals in the body is fairly well understood. Trace minerals are needed daily in smaller amounts. These include copper, iodine, zinc, iron, manganese, and more than 15 others. Table 9.8 outlines the minerals, their RDA's, primary functions in the body, and major dietary sources.

Vitamins and Minerals Supplementation.   With a good balanced diet, all the vitamins and minerals required by the body can be obtained. However with busy schedules, diets often contain many processed and refined foods causing most individuals to not receive their RDA. Therefore, it seems prudent to take one multi-vitamin plus mineral supplement per day with not more than 100 % of the RDA as a good insurance policy against any possible deficiency. Megadoses of vitamins and minerals should be avoided as they will not improve health or athletic performance but can lead to many health problems (see Table 9.9).

TABLE 9.8

RDA Amounts for Minerals, Their Functions and Major Source

MINERAL	ADULT RDA	PRIMARY BODY FUNCTION	MAJOR SOURCE
Calcium	800 mg	Needed for formation of teeth and bones, nerve impulse transmission	Dairy products, broccoli, sesame seeds, salmon
Phosphorus	800 mg	Needed for formation of teeth and bones, maintain the acid-base balance of the blood	Dairy products, fish, eggs, whole wheat, beans, nuts
Sodium	2.5 gm	Needed for nerve impulse transmission, water balance, acid-base balance of the blood	Common salt, processed foods
Chlorine	2.0 gm	Forms stomach acid, maintains acid-base balance of the blood	Common salt, processed foods
Potassium	2.5 gm	Functions with sodium for nerve impulse transmission, water balance, acid-base balance of the blood	Abundant in most plant and animal foods
Magnesium	350 mg	Part of enzymes that metabolizes nutrients into energy	Soybeans, peanuts, walnuts, wheat germ, spinach
Sulfur	Not Known	Important part of all proteins	Meats, dairy products, beans
Iron	10 mg	Important part of hemoglobin	Beans, whole grains, eggs, liver
Copper	2 mg	Helps to combine iron with hemoglobin	Beans, nuts, fish
Fluorine	2 mg	Helps prevent tooth decay	Most drinking water, fish
Zinc	15 mg	Associated with insulin relative to carbohydrate metabolism	Soybeans, wheat germ, nuts
Iodine	15 mg	Important part of thyroid hormones	Vegetables, fish, dairy products
Silicon Vanadium Tin Nickel Selenium Manganese Molybdenum Cobalt Chromium	Not known but needed in small amounts	Needed in very small quantities for a variety of functions	Vegetables, wheat germ, grains, dairy products, and fish are the best overall sources

TABLE 9.9

Health Problems Associated with Megadoses of Vitamins

Ϋ Vitamin C	**Gout, Hemolytic Anemia, and Diarrhea**
Ϋ Vitamin B6	**Liver Disease and Nerve Damage**
Ϋ Vitamin B2	**Vision Impaired**
Ϋ Vitamin E	**Headaches, Fatigue, Blurred Vision, Muscular Weakness, and Gastrointestinal Disturbance**
Ϋ Vitamin A	**Nervous System Damage**
Ϋ Vitamin D	**Kidney Damage**

Natural vs. Synthetic.   The human body identifies vitamins and minerals by their chemical structures. Vitamin C has the same chemical structure whether it is derived from citrus fruit or synthetically made. People who promote "natural" or "health" food are doing so to capitalize on the public's interest in good health and willingness to pay more for it. Today, the supplement and health food industry is a billion-dollar business.

7. Water

The human body is made up of 50 to 70 % water making it extremely important to consume enough fluids daily. The body requires approximately eight glasses of water a day. Some of this liquid may be in the form of milk, soup, and juices, but it is best to get most of the fluids from water. One should drink enough fluids daily so that the urine becomes so diluted it is almost clear and colorless at least once during a 24 hour period.

Most fruits and vegetables are also good sources of water such as tomatoes, eggplant, cauliflower, lettuce, strawberries, and watermelon which consist of over 90 % water. The amount of water ingested daily will vary from eight glasses in a normal day to an equivalent of 10 to 12 on hot or exercising days when extra fluid is lost through perspiration.

Caffeine-containing drinks such as coffee, tea, and colas should be avoided or the amount decreased significantly. Caffeine is a diuretic as well as a central nervous system stimulant and can cause irregular heartbeats, formation of acid-pepsin digestive juices in the stomach, and fibrocystic breast disease in women. A cup of coffee contains 100 to 150 milligrams of caffeine; a cup of tea contains 50 to 75 milligrams; a 12-ounce cola contains 60 milligrams; and a cup of hot chocolate contains about 50 milligrams.

Purchasing and Preparing Food

Purchasing and preparing food are two important aspects of good nutrition. When grocery shopping, purchase most of the food from the fresh produce and fresh frozen section. Most of the daily caloric intake should come from fresh, whole foods, not processed ones. Buy fruits and vegetables that are in season, and plan meals around them. Recipes for delicious meatless dishes using vegetables and rice are plentiful. By shopping at roadside markets with locally grown produce, the nutritional value of foods is usually superior.

When choosing meats, poultry and fish should be first choices. Many grocery stores have fresh fish, both freshwater and ocean varieties. Chicken is always an economical buy, and there are numerous appetizing ways to prepare it. Both chicken and fish, properly prepared, provide a dish high in protein and low in fat.

Become a label reader and do not be confused by the words fortified and enriched. Most of these products have been stripped of most of their vitamins and minerals and then had a handful of nutrients added back. Emphasize brown rice over white rice, since brown rice has not been refined. Buy cereals with as much of the whole grain in them as possible. Read the nutritional breakdown on the box and notice the fiber content. Look at the complex carbohydrate and sugar content keeping it high in complex carbohydrates and low in sugar. Notice bread ingredients because some breads claim to contain whole-wheat flour and look brown due to the caramel coloring. Look for ingredients such as wheat bran, wheat germ, wheat berries, or perhaps a combination of grains such as rye, barley, or oats. The coarser and grainier the bread, the better.

When preparing fruits and vegetables, peel and trim as little as possible. Raw fruits and vegetables keep vitamins and minerals. If cooking the fruit or vegetable, scrub it immediately prior to cooking but do not soak it. Cook as briefly as possible and serve immediately. Microwave cooking of vegetables is also good because little water is used.

Skin the poultry and bake, broil, stir fry, or boil it, then use it in salads and casseroles. Fish is also healthier if it is broiled or baked. Experiment with various recipes and seasonings to find something just as tasty as fried chicken or fish.

Several other practices will help to maintain a healthful eating style. Use vegetable oils instead of solid shortening not only in frying but also in baking. For instance, if a cookie recipe calls for one cup of butter, use half oil and half margarine. When using eggs in cookies, cakes, or breads, leave out the yolks and substitute two whites for each whole egg. This cuts down on cholesterol intake since the egg yolk is a high source of cholesterol. When whole eggs must be used, consider egg substitutes.

The sugar called for in most recipes can be cut by one-third to one-half and not affect the taste or texture. To increase fiber in the diet, bran and wheat germ can be mixed in or sprinkled on most foods without affecting taste or texture. Sprinkling a tablespoon of wheat germ on cereal in the morning or on vegetables and casseroles. Bran is great in muffins and breads.

The Basic Food Groups

The United States Department of Agriculture in 1992 revised the traditional Basic Four food groups into the following seven categories as outlined below:

1. **Breads, cereals and other grain products.** The new suggested servings are 6-11 per day of breads, cereals and other grain products. This group of foods is considered the most important and the foundation of a healthy diet. A serving would be one slice of bread, one ounce of cereal, or one-half cup of rice. The best choices in this group would be 100 percent whole-grain breads, brown rice, and cereals as described earlier. Acceptable foods are crackers, white rice, pizza, macaroni, and waffles.

2. **Fruit.** The new suggested servings of fruit are 2-4 servings per day. An example of a serving would be one whole fruit (banana, orange, apple, etc.), one-half a cantaloupe or grapefruit, one slice of watermelon, one-half cup of most berries, and six ounces of fruit juice.

3. **Vegetables.** The new suggested servings of vegetables are 3-5 servings per day. A serving size would be a half-cup of most vegetables. The best of this group would be raw vegetables. Cooked vegetables and canned vegetables are acceptable.

4. **Protein Group 1.** A serving would be two to three ounces of meat, two egg whites, one-half cup of legumes, one-half cup of nuts or seeds, or two tablespoons of peanut butter. The best sources for this group would be roasted chicken or turkey, fish, egg whites, and legumes such as lentils, peas, or beans. Lean beef, seeds, and nuts are acceptable. Because of the high fat content, the following foods in this group are to be avoided: fish sticks, commercially prepared chicken, egg and cheese dishes, bacon, cold cuts, and frankfurters.

5. **Protein Group 2.** A serving would be eight ounces of milk, eight ounces of yogurt, one-half cup of cottage cheese, or one-inch cube of cheese. The best of this group would be skim milk, low-fat yogurt, and low-fat cheese. Two- percent milk and ice milk are acceptable. The simplest way to achieve the requirement in this group is to have one or two glasses of skim milk a day. That is all an adult needs. Any more from this group adds unnecessary fat to the diet.

6. **Fats (unsaturated).** Pure fats from oils should be limited in the diet as much as possible with a maximum of 30 to 40 grams of unsaturated fats allowed. One tablespoon of fat (such as butter, margarine, vegetable oil, or salad dressing) has 14 grams of fat.

7. **Sweets and Alcoholic Beverages.** Sweets in the form of candies, pastries, desserts and alcoholic beverages should be avoided or limited in the diet.

Eat Well-Balanced Meals

People today tend to skip meals, especially breakfast. Lack of time is the usual excuse, but is important to eat balanced meals each day. Decide how many calories are needed to consume each day to lose, maintain, or gain weight, depending on your situation. Determine what 30 percent of those calories would be and let that be the amount for each meal: breakfast, lunch, and dinner. The remaining 10 percent may be divided up into midmorning, midafternoon, or evening snacks. Do not wait until evening to consume all the calories since it appears that food eaten after 2:00 P.M. tends to be stored as fat. Eating most of the calories before 2:00 P.M. gives the body more hours to digest and burn them.

The following examples are suggestions for what one-day of meals, including snacks, might look like:

Breakfast

1. 6 ounces orange juice, or 1 whole orange

2. Bowl of high-fiber cereal

3. Banana

4. 1 slice 100% whole-wheat toast with jelly, no margarine or butter

5. 8 ounces skim milk

Lunch

1. Sandwich of tuna or chicken on 100% whole-wheat bread with lettuce or sprouts

2. Carrot and celery sticks

3. Fresh fruit (apple, pear, orange, etc.)

4. 8 ounces skim milk

   (Peanut butter could be substituted for meat, and a tossed salad could replace carrots and celery. Yogurt is also good and is easy to make at home without all the added sugar of the ones made commercially.)

Dinner

1. 4-6 ounces baked fish or chicken

2. 1 medium baked potato, sweet potato, or brown rice

3. 1/2-3/4 cup fresh steamed broccoli or spinach salad

4. 1/2-3/4 cup carrots or corn

5. Water

6. Piece of fresh fruit or fruit salad

Snacks and Desserts

Snacks and desserts can have a place in the daily diet. Some suggestions for snacks include raw vegetables, fruit, cold cereals, and popcorn (limit salt and butter). Raw nuts are good but only in limited quantities because they are high in fat and calories. Dried fruits can satisfy a craving for something sweet. Most grocery stores have raisins, dried apples, bananas, prunes, and so on. Here again the amount should be limited because of the high sugar content. Desserts don't have to be eliminated from the diet, only modified or decreased. Cutting a piece of dessert in half each day may save 100 to 300 calories.

Guidelines for Controlling Caloric Intake

1. Sit down to eat at the kitchen or dining room table. Get used to eating in one place. Don't eat in front of the TV, at your desk, at a movie, or at an athletic contest. This alone could eliminate several hundred calories a day.

2. Eat only at specific times. This will cut down on unnecessary snacking.

3. Use a smaller plate when possible. A small plate that is filled can look like it has more food than a large with the same amount of food.

4. Eat slowly. Not only is this better for digestion, but the allows the appestat signals to reach the brain when the stomach is full, before over eating occurs.

5. Carry a calorie chart to help learn the approximate number of calories in most of the foods consume. Small books are available at grocery stores and bookstores that fit into a woman's purse or a man's suit coat pocket.

6. Don't use food as a reward.

7. Make a shopping list when buying food and stick to it. Don't buy on impulse.

8. Don't grocery shop when hungry.

9. Store food at home out of sight. Don't keep cookies and snacks on the counter.

10. Clear the table immediately after a meal.

11. Whenever possible, eat raw, unprocessed foods. For example, a potato costs about 12 cents per pound and contains no fat and little sodium. Potato chips, on the other hand, cost two dollars per pound and contain 65% fat and 100 times the sodium of a potato.

12. Little things mean a lot. One snack or drink every day could add up to 36,500 calories in one year which equals about 10 pounds of body fat. A daily half-hour walk could burn 67,700 calories in one year or 18 pounds of fat.

Summary

Health is very dependent upon the food and fluids consumed each day. The seven basic nutrients are carbohydrates, fiber, fats, proteins, vitamins, minerals, and water. Complex carbohydrates should be increased in one's diet and simple sugars decreased. The overall dietary intake of carbohydrates should be increased from the present 45 to 60 % for good health. Fats and cholesterol should be decreased in the diet. High-protein diets are undesirable, and it is possible that a highly varied, semi-vegetarian diet can serve one best in most situations. A daily multivitamin plus mineral supplements to one's regular diet make a good insurance policy against nutritional deficiency. The basic food groups include: 6-11 servings of breads, cereals and grains; 2-4 servings of fruit; 3-5 servings of vegetables; 2-3 servings of protein group 1 (legumes, nuts, chicken, fish, etc.); and 2-3 servings of protein group 2 (dairy products). Fats, sweets, and alcoholic beverages should to be avoided or limited.

Questions

1. What are the two overall functions of nutrients?

2. List the seven basic nutrients needed for maintaining health of the body.

3. What is the primary function of carbohydrates?

4. Which body cells use only carbohydrates for energy?

5. A restricted carbohydrate diet on high energy demands could have what affect on protein in the body?

6. Compare the time of digestion between simple sugars and complex carbohydrates (starches).

7. What is the function of insulin from the pancreas?

8. What is fiber, and what is its function in the body?

9. List the primary sources of fiber and how much fiber one should have a day.

10. What is the function of fats in the body?

11. Compare saturated and unsaturated fats?

12. What is the function of protein in the body?

13. What happens to excess protein in the body?

14. What are the functions of vitamins; distinguish between fat-soluble and water-soluble vitamins.

15. What are the current seven basic food groups?

10

Managing Back Injuries and Stress

Learning Objectives

This chapter examines ways in which to manage back injuries and stress to improve one's health fitness. Even those who are physically fit still encounter stressors every day, and how they handle these will increase their health fitness or distract from it. When you complete this chapter you should:

1. Know the types of back problems and some recommendations for preventing them.

2. Be able to describe guidelines and exercises that promote a healthy back.

3. Understand the role that perception plays in reducing stress.

4. Know examples of four groups of stressors.

5. Recognize the immediate and long-term affects of stress on the body.

6. Understand the immediate and long-term benefits of physical activity relative to stress.

7. Describe at least six behaviors that can assist with effective stress management.

Key Terms

Back Strain/Sprain

Deep Breathing

Muscle Relaxation

Meditation

Posture

Slipped Disc

Stress

Stressors

Types of Back Problems

An estimated 80% of the American adult population has suffered from low-back problems at one time or another. Back problems account for more lost man-hours than any other occupational injury and consequently accounts for a significant portion of mounting health costs. Americans spend more than $5 billion a year for tests and treatments from orthopedic physicians, osteopaths, physical therapists, and chiropractors.

Back problems can take several forms. The pain may result from pathological problems such as arthritis, rheumatism, a cancerous tumor growing on the spine, osteoporosis, or infection. It may result from a fall, a blow to the spinal column, or pregnancy. But the vast majority (over 85%) of back problems are not caused by any of these reasons. Instead, they are due to muscle deficiency caused by an inappropriate lifestyle. Two of the most common back problems are back strain/sprain and slipped disc.

1. Back Strain/Sprain

The most common back problem is a back strain/sprain which usually is the result of working hard in the garden, moving the furniture, or undertaking some other physical activity in which the back muscles were required to do more than they were capable of doing at that moment. The injured muscles or tendons cause inflammation, fluid build-up, muscle spasms, and painful pressure on the nerve endings. The muscle spasms constrict the blood vessels and cause further complications by not allowing as much blood with oxygen and nutrients to reach the area or waste products to be carried off.

The immediate treatment for a back strain/sprain is rest, aspirin or ibuprofen to reduce the inflammation, and perhaps a muscle relaxant prescribed by a physician to relax the muscle spasm. Usually in a few days the pain will be gone, but it will return if a preventive program is not followed.

2. Slipped Disc

Although a slipped disc may be the most talked about back ailment, it accounts for only five percent of back problems. Back strains/sprains account for the other 95%; however, if a back strain/sprain is severe enough, it can mimic a slipped disc.

A slipped disc is technically not slipped. The cartilage that serves as a cushion between the vertebrae is known as a disc. When excess pressure is placed upon the disc, it may herniate (rupture). When this occurs, some of the jelly like center of the disc bulges out of the disc and pushes against the nerves in the spinal cord. These nerves are extremely sensitive, and the least pressure against them can trigger excruciating pain.

A long history of repeated attacks of muscle strain/sprain can begin the process of disc deterioration. Initially the disc deterioration may cause no symptoms and go unnoticed. This process may eventually lead to a ruptured disc.

Symptoms of a Ruptured Disc.   Although a ruptured disc is often difficult to diagnose, two symptoms are usually present. The first system is pain going down one or both of the legs called sciatica. A rupture disc is most commonly located in the lower back where it places pressure on the sciatic nerve. If there is no pain down the leg, the problem is probably not from a disc rupture. The second symptom strongly indicative of a ruptured disc is pain in the lower back when lifting one straight leg off the floor while lying flat. If the straight leg can be lifted at least 70 to 80 degrees without causing pain in the low back, there is probably no ruptured disc. (This test should never be performed when significant back spasms are occurring. If, on the other hand, the straight leg can be lifted only 20 or 30 degrees before pain is felt this is an indication of a possible ruptured disc and a physician should be seen.

Treatment for a Ruptured Disc.   Very often a ruptured disc requires surgery. However, unless there is a change in lifestyle, the problem will return following the recovery from surgery. It is possible for a ruptured disc to heal in time without surgery and not recur if there is a significant lifestyle change.

Decreasing Back Injuries

Three of the most common reasons for back problems are 1) weak abdominals and spinal muscles with poor hip, spine, and leg flexibility, 2) poor postural habits and body mechanics, and 3) overfatness. By incorporating specific flexibility and strengthening exercises into a regular fitness program, back injuries can be decreased.

Strength Exercises. Four specific muscle groups that support the vertebral column, the pelvis, and the abdomen must have adequate strength to prevent back problems.

Back Muscles support the entire vertebral column and must be strong. If they are weak, heavy work will strain them and cause back pain.

Abdominal Muscles must be strong to keep the pelvis tilted up and prevent it from rotating down (giving a potbelly or swayback appearance). When the pelvis is rotated down, undesirable pressure is placed on the vertebrae of the lower back, on the discs, and on the nerves that branch from the spinal cord to the muscles in the low-back area and legs.

Gluteal Muscles are also important for stabilizing the pelvis and assist in preventing it from rotating down in front. They are attached to the back of the pelvis and the upper part of the back of the thighs.

Iliopsoas Muscles must be strong and flexible. This muscle group has as its upper attachment the inside of the vertebral column of the lower back and runs through the pelvic girdle and is attached to the front of the upper thigh. This muscle group is important in hip flexion and also supports and stabilizes the lower part of the back.

Flexibility.   Three key muscle groups which keep the body upright are always mildly contracting during sitting, standing, walking, or jogging. These muscle groups are the 1) back, 2) hamstrings, and 3) iliopsoas. Because they are mildly contracting for much of the day, they will develop tightness if specific types of stretching exercises are not performed.

Guidelines for Back Exercise

1. Learn to distinguish between pain and discomfort. Pain while exercising is a signal to stop. Some discomfort is to be expected and will go away shortly after the exercise is finished. Exercise should not be done when pain is present.

2. Progression will be slow if recovering from a back problem, allow three to six months to achieve the strength and flexibility needed for the back to resist strain/sprain. In fact, noticeable improvement in a month or two is not uncommon.

3. Exercise every day when first beginning a back exercise program since the number of repetitions will be so few. After achieving satisfactory levels of strength and flexibility, the exercises need to be performed only three days per week for maintenance.

4. A warm-up should be performed before exercising. If there is back discomfort, taking a short walk, a warm bath, or shower before doing the back exercises will help.

Specific Exercises for Preventing Back Problems

1. **Curl-up**. Lie on the back with knees bent. Raise the head, contract the stomach muscles, and reach up until touching the knees, then lower back down to the starting position. Begin by doing five repetitions, and

add one or two repetitions each week. The goal is to do 50 without stopping. After a month of performing 50 curl-ups daily, then gradually replace them with sit-ups. For example, do 45 curl-ups and five sit-ups for two weeks. Then progress to 40 curl-ups and 10 sit-ups for two weeks. If no pain occurs during the sit-ups, continue to increase them until all 50 are sit-ups. *(Sit-ups should be avoided when recovering from a back problem.)*

2. **Pelvic raises.** Lie on the back with knees flexed, feet flat on the floor, and arms at the sides resting on the floor. Raise up the pelvis as high as possible by contracting the gluteal muscles, and then lower the pelvis back to the floor. Begin with five and add one or two each week until 50 can be performed without stopping.

3. **Side leg raises**. Lie on one side of the body with the head resting in one hand with the arm bent and elbow on the floor. Raise the upper leg as high as possible keeping the knee straight and hip moving forward. Move the leg slowly up and down. Begin with five on each side the first week and add one or two each week until performing 50 without stopping.

4. **Knees to chest or low-back stretch**. Lie on the back with legs straight. Raise right knee and hug it to the chest with the hands. Repeat using the left knee and then with both knees together. Begin by doing three a day, and then add one per week until able to do 20. When just recovering from a back problem, this exercise will cause some discomfort in the back muscles. However, as the back heals and flexibility increases, the discomfort will disappear. As flexibility increases, reduce the repetitions to 10, and add the next exercise.

5. **Hamstring stretch**. Sit on the floor with one leg straight and the other bent. Bend forward at the waist over the stretched leg reaching the fingers toward the toes. Switch the legs and repeat. Perform this exercise after mastering the low-back stretch with no discomfort. Start with three repetitions holding for 10 seconds, and add one per week until 10 repetitions can be done.

6. **Quadriceps stretch**. Lay on the side of the body with both legs bent. Grab the top leg at the ankle and gently pull back stretching the quadriceps. Switch to the other side and repeat with the other leg. Start with three repetitions, hold for 10 seconds, and add one per week up to 10 repetitions.

7. **Back extensions.** Lie face down on the floor with the elbows by the chest, forearms on the floor and hands under the chin. Gently raise the trunk by contracting the back muscles. Once the chest is as far off the ground as possible, lower slowly. Perform this exercise only after back pain has stopped. Begin by doing two, and add one each week until performing 14.

8. **Hanging**. Whether upright or inverted, hang for 30 to 60 seconds each day. Hanging helps the vertebral column to elongate, which reduces the pressure on each vertebral disc. *(Do not hang inverted if you have heart disease or high blood pressure).*

Other Lifestyle Factors

Several additional factors are crucial in preventing back problems. These include posture, lifting habits, weight reduction, and stress management.

1. Posture

Good posture is a habit, and habits are formed by constant repetition. Good posture results from strong muscles, and strong muscles require a muscle-development exercise program. By developing good positive habits and strong muscles, most back problems can be avoided.

If there was one word that best conveys the mental image of good posture and body alignment, it would be "tall" (Think tall). Adherence to the following three practices will lead to good posture:

1. Hold your head up high and tall. Feel as though you are trying to touch the ceiling with the top of your head.

2. Raise your chest, rib cage, and shoulders.

3. Tilt your pelvis backward by contracting your abdominal muscles and your gluteal muscles.

In addition to practicing the above postural habits, try to remember to place one foot on a stool, chair, or other object when standing for long periods. Also, alternate the feet to allow shifts of weight while keeping the back from excessively swaying (lordosis). When sitting place the feet, with knees bent, on a stool or a footrest. The chair should have a good, firm seat with a cushion. Do not slump. The same advice applies while driving a car. The knees should be bent. Move the seat of the car as close to the steering wheel as feasible to prevent swayback. It is also helpful on long trips to stop at rest areas that furnish tables or benches that may be used for back exercises and stretching.

2. Lifting Habits

Improper lifting of heavy objects or carrying heavy objects can place considerable pressure on the vertebral column and lead to back problems. Thus, proper lifting technique is extremely important. When lifting objects from the floor, the knees should be bent so that the lifting is done with the legs rather than the back. When placing groceries into the trunk of a car, raising windows, hanging up clothes, cooking, making a bed, or reaching for objects, lift with the object as close to the body as possible. When lifting objects while turning, since there is less stability and support of the vertebrae, the feet and entire body should be rotated as a unit.

3. Weight Reduction

Excess body fat, in addition to placing harmful pressures on the cardiovascular system and other systems of the body, places undesirable stress on the vertebral column. An extra 20 or 30 pounds often carried on the abdomen, places considerable strain upon the back and can lead to back pain.

4. Stress Management

If a person is under constant stress, all the muscles in the body become tense. The slightest jar can cause a strain/sprain in a back muscle. Persons with low-back pain often have a recurrence when they are involved in a stressful situation. It is important to effectively manage stress.

Stress

Hans Selye, who is considered by most experts as the leading authority in the study of stress, defines stress as the "response of the body to any demand made upon it." Stress is the body's response to counteract any stressor. It is the body's way to maintain homeostasis, or internal environment. Whenever the body's equilibrium is upset, the body attempts to return it to homeostasis.

There are four classifications of stressors:

1. **Social stressors:** relationships with family, friends, and co-workers; financial concerns; changes in one's life; and etc.

2. **Environmental stressors**: heat, humidity, cold, noise, heavy traffic, overcrowding, pollution, and etc.

3. **Physical stressors**: illnesses, injuries, lack of sleep, caffeine or nicotine, and etc.

4. **Psychological stressors**: fear of losing one's job, worry over grades in school, loneliness, anger, guilt, lack of self-esteem, frustration, and etc.

Perception of Stressors

The perception of the significance of a stressor often determines the body's response to it. More often than not, stressors are intangible, symbolic, and subject to each person's interpretation. If the mind perceives the stressor as significant or "real", the body's stress response will be set in motion, but this will not occur if the mind perceives the stressor as insignificant.

Perception is influenced by several factors. First, the spiritual dimension of the person influences perception. If the person has a positive belief system, faith, and self-confidence, the person is less likely to perceive the stressor as a serious threat, and the stress response will be minimal. On the other hand, if the person's spiritual dimension is weak and there is fear and worry or a lack of self confidence, the stressor probably will be viewed in a negative light, and the stress response will be greater.

Second, perception is influenced by the intellectual ability. If the stressor is a challenge to the intellect such as when a student has a paper to write, or an individual has a problem to solve, but the person has previous knowledge on how to complete the task, the stress response is usually minimal. On the other hand, if one does not possess the knowledge necessary to write the paper or solve the problem, the stressor will be perceived negatively, and the stress response will be greater.

The tendency too often is to think negatively, dwell on problems, and focus on aches and pains. Dwelling too long on negative thoughts can cause the subconscious mind to begin to take control and paralyze the conscious mind. The body will react, and sickness and depression can result. When negative thoughts occur, a healthy response is to capture them, control them, evaluate them, and change them into positive thoughts. Positive statements should be repeated many times a day until they are driven into the subconscious mind. Then the body will begin to react according to the positive thoughts. Controlling ones thoughts is imperative for good health. Negative thoughts can lead to feelings of defeat, anxiety, distress, and depression; and these emotions can spark many health problems.

Symptoms of Stress

Often the results of poorly handled stress are obvious. Symptoms include feeling tense, having cold and clammy hands, and sleeping difficulties (see Table 10.1). Recognizing when the body is not responding satisfactorily to stress is the first step to overcoming poor stress management.

If a person does not successfully manage the stress and positively adapt to the stressors, the stress response will lead to health problems. Selye believes that when the body is under a chronic state of stress, it is in a constant state of readiness to respond, and almost any disease may develop. Some of these diseases are shown in Table 10.2.

TABLE 10.1

Immediate Symptoms of Stress

✗ Pounding of the heart	✗ Depression
✗ Tension	✗ Trembling
✗ Insomnia	✗ Pain in the neck
✗ Fatigue	✗ Irritability
✗ Dizziness	✗ Nervous tic
✗ Grinding of the teeth	✗ Sweating
✗ Pain in the low back	✗ Lack of concentration

TABLE 10.2

Symptoms of Long Term Stress

➡ High Blood Pressure	➡ Ulcers
➡ Chronic Low Back Pain	➡ Chronic Neck and Shoulder Pain
➡ Heart Disease	➡ Stroke
➡ Migraine Headaches	➡ Tension Headaches
➡ Arthritis	➡ Chronic Fatigue
➡ Depression	➡ Insomnia

Role of Physical Activity in Stress Management

According to the American National Institute of Mental Health, physical activity has both immediate as well as long term benefits. When engaged in physical activity, *immediate benefits* include:

1. Physical activity takes the mind off the problem; it is difficult to worry when focused on a physical activity. When the activity is over, the thought processes become more efficient.

2. Physical activity is invigorating, and when it is over a relaxed feeling occurs reducing anxiety. Muscles tension is relieved, and an improved feeling of self worth occurs.

3. Physical activity quickens the breakdown of the chemicals released during the stress response, and the body returns to homeostasis faster.

4. Physical activity releases endorphins, the body's natural tranquilizers.

Many business executives wisely conduct an exercise program at 5:00 or 6:00 P.M., after a busy day of work, to help relieve the tensions of the day. Exercise has been shown to be beneficial in the treatment of depression. Therapists prescribe aerobic exercise as a therapeutic mode to treat depressed patients. However, not all physical activity will necessarily provide the same benefits. Participation in sports activities such as golf, tennis, and racquetball may not assist the body in relieving stress the same way that aerobic activities do such as jogging, walking, and cycling. In competitive sports, where there is pressure to perform well or to win, stress may be increased rather than decreased.

Long-term benefits of physical activity include:

1. Regular exercise improves the health and fitness of the body making it better able to withstand the stressors.

2. Regular exercise promotes a positive self-concept and improves self-confidence.

Effective Stress Management

Since stress is inescapable in today's society, it is vital that one learns how to successfully manage it. While it is beyond the scope of this book to thoroughly discuss stress management techniques, the following are some general guidelines:

1. **Plan Ahead.** Most stress is due to lack of planning ahead or becoming over committed. Effective planning involves setting realistic goals. Do not plan so much for a day, week, or month that it is impossible to

accomplish it all. Also, prioritize tasks to be done so that time will not be wasted on unimportant activities. Develop a list of things to be done, prioritize them, and then set a specific time aside for doing each one.

2. **Replace fear with positive thinking.** Fear of physical danger can be beneficial by inducing the extra physical energy needed to escape. Too often, however, fear is not of a physical danger but of failure. This type of fear is harmful to the body because rather than causing an active response, the common response is lethargy. Attempt to replace fears with positive thoughts about oneself and develop a positive self-concept.

3. **Relaxation.** Like aerobic exercise, relaxation techniques can reduce the potential harmful side effects of stress. There are several relaxation procedures.

   Deep Breathing Exercises are simple ways to help reduce stress. Take deep breaths and let the air out slowly. With each breath feel the stress going from the body. Repeat 5 to 10 times.

   Muscle Relaxation is another way to reduce stress. Lie down in a quiet and peaceful place and alternately contract and relax the muscles beginning at the feet and working up to the head.

   Meditation is another method to aid in relaxation. Studies demonstrate that, with practice, after 5 to 10 minutes of meditation, heart rate and metabolism have slowed down indicating its effectiveness in stress management. There are several meditation techniques. One method is to sit quietly, eyes closed, and repeat a simple word or phrase. Another is to take a mental vacation and picture yourself at some location that is especially peaceful to you.

4. **Avoid Caffeine and Nicotine.** When feeling under tension or stress, avoid caffeine and nicotine. Both substances are stimulants and add to the stress reactions of the body.

5. **Be a Part of a Supportive Social Group.** Numerous studies are confirming the need to be a part of a positive social group to reduce stress and related disease. Intimacy, openness, and sharing all help to reduce stress and improve health.

Summary

Effectiveness in overcoming back pain is through muscle stretching and increasing flexibility in the back, abdomen, hip, and thighs. With 80 % of the adult population suffering from back pain, it is important that appropriate exercises are performed regularly. Stress management involves the total person. Regular physical activity, prioritized time management, positive thinking, relaxation and meditation, avoidance of stimulants, and positive social interaction all contribute to effectively assist in stress management.

Questions

1. List four lifestyle factors that are crucial for preventing future back problems.

2. List the exercises that are prescribed for preventing back problems.

3. Distinguish between back sprain/strain and herniated disc.

4. Which muscle groups need to be strengthened for maintaining a healthy back?

5. Define stress and homeostasis.

6. What are four groups of stressors?

7. Discuss the role that perception can have on stress.

8. What are some immediate and long-range symptoms of poor stress management?

9. What are some immediate and long-term benefits of physical activity relative to stress management?

10. List five recommendations in addition to regular aerobic exercise for effective stress management.

Laboratories

Contents

Medical Clearance

Name _______________________________ Section ____________ Date ____________

There are many health benefits that are associated with regular exercise. However, before increasing your physical activity, medical clearance is the first step to take. While measures will be taken to insure your safety, this class does involve physical activity and you must assume the risk of an injury or accident. While exercise is safe for most apparently healthy individuals, the reaction of the body can't always be totally predicted. Any exercise program can be a risk; but the hazards of sedentary living are far greater than most risks associated with exercise.

Before starting any exercise program, please read carefully the following statements to determine if any situation applies to you that may limit your physical activity. This procedure has been designed to identify contraindications that would warrant the advice of a physician concerning the type of activity most suitable. Therefore, according to the guidelines of the American College of Sports Medicine, before participating in the exercise portion of this class:

1. Each student should notify the instructor if there are any medical problems or physical disability that would limit full participation in class.

2. Students over the age of 40 (males) or 50 (females) should consult with their physician to determine if they are "cleared" to fully participate in the class. If "clearance" is not received, a prescription of exercise modifications is essential for safe participation in an exercise program.

3. Students of any age with any major coronary risk factors and/or symptoms suggestive of cardiopulmonary or other disorder listed below should consult with their physician to determine if they are "cleared" to fully participate in the class.

    a. Diagnosed high blood pressure.

    b. High serum cholesterol.

    c. Cigarette smoking.

    d. Diabetes.

    e. Family history of heart disease in parents or siblings prior to age 55.

    f. Chest pain at rest or during exertion.

    g. Unaccustomed shortness of breath or shortness of breath with mild exertion.

    h. Dizziness.

    i. Labored breathing.

    j. Heart palpitations and/or arrhythmias.

    k. Known heart murmur.

    l. Joint, muscle, or other orthopedic problems

4. Students should carry health insurance. The university is not financially responsible for any medical care that may result from injuries or accidents occurring in this course.

I acknowledge that I have read this form in its entirety and that I understand the risks involved in beginning an exercise program. I accept the responsibility of being fully "cleared" to participate in an exercise program and agree to provide information to my instructor concerning any situation that may exist that would prohibit or limit my full participation in this class.

Name (print) _________________________________ ID __ __ __ - __ __ - __ __ __ __

Signature _______________________________________ Date_________________

Age __________ Gender (circle one)   Male   Female

Class and section _______________________________ Time _______________

Instructor ____

An Estimate of Risk of Having a Heart Attack

Name _________________________________ Section _____________ Date _____________

Evaluate your chances of having a heart attack by reading each section and recording your score.

1. *Heredity* (consider parents, grandparents, brothers, sisters)

    a. No relative under 75 with heart disease   0

    b. One relative between 65 and 75 with heart disease   1

    c. Two relatives between 65 and 75 with heart disease   2

    d. One or two relatives 55 to 65 with heart disease   4

    e. One or two relatives under 55 with heart disease   6

    f. Three or more relatives under 55 with heart disease   8

    SCORE _________

2. *Age*

    a. Under age 30   0

    b. 31-40   2

    c. 41-50   4

    d. 51-60   6

    e. 61 plus   8

    SCORE _________

3. *Sex*

    a. Female under 40   0

    b. Female 40-50   1

    c. Female over 50   3

    d. Male under 50   3

    e. Male 50 plus   5

    SCORE _________

4. *Blood Pressure**

    a. Systolic under 100   0

    b. Systolic 100-120   1

    c. Systolic 121-140   2

    d. Systolic 141-159   4

    e. Systolic 160-180   6

    f. Systolic over 180   8

    SCORE _________

5. *Smoking*

    a. Nonsmoker   0
    b. 10 or fewer cigarettes a day   2
    c. 10-20 or fewer cigarettes a day   4
    d. 21-40 or fewer cigarettes a day   8
    e. Over 40 cigarettes a day   10

SCORE ________

6. *Obesity and Excess Weight*

    a. Within 5 Lbs normal weight   0
    b. 6-20 Lbs over   1
    c. 21-50 Lbs over   4
    d. 51-80 Lbs over   7
    e. More than 80 Lbs over   10

SCORE ________

7. *Exercise and Fitness***

Can walk/jog 1.5 miles:

Men	**Women**	
Under 10 minutes	Under 11 minutes	0
10-12 minutes	11-13 minutes	1
12-14 minutes	13-16 minutes	2
14-18 minutes	16-20 minutes	4
18 plus minutes	20 plus minutes	8

SCORE ________

8. *Cholesterol****

    a. Below 160   0
    b. 160-200   1
    c. 200-240   3
    d. 240-300   6
    e. 300 plus   10

SCORE ________

9. *Emotional Stress*

    a. *I always* feel relaxed, *nothing* ever bothers me.   0

    b. I occasionally get uptight.   2

    c. I get uptight several times per week.   4

    d. I get uptight usually once a day or more.   6

    e. I am usually in a hurry; there is just not enough time to get things done. People sometimes get in my way. To accomplish my goals, I often feel tense.   8

SCORE ________

TOTAL SCORE ________

Risk Factor Score

6-13	well below average
14-22	below average
23-29	average
30-36	above average
37-44	high level
45 plus	very high

*If you don't know, score 2.

***Do not attempt this run* unless you are in good physical condition and have been cleared by a physician. Score 8 if you do not meet this criterion.

***If you don't know your cholesterol, score 4. If you emphasize a low-fat diet, then score 2.

This chart is only an estimate of risk factors and should not be considered an actual prediction of either having or not having a heart attack; however, if your risk is high, you should take steps to lower it.

Evaluating Body Composition

Name _________________________________ Section ____________ Date ____________

Introduction

Excessive body fat is associated with a variety of diseases. Determining body fat allows students to compare their own body composition to that of an acceptable level for health.

Purpose

1. To determine body composition by using skinfold measurements.

2. To help the students evaluate and rate their own body fat composition.

3. To learn and understand what is an acceptable body fat percentage.

Procedure

1. Students will wear appropriate clothing conducive to testing as recommended by the instructor.

2. Appointed testers will measure skinfold sites with calipers.

3. Using the appropriate tables, students will determine and record their estimated body fat percentages.

4. Students will determine their fat and fat free weight in pounds.

Results and Discussion

1. Record your weight without shoes and clothes to the nearest pound __________ lbs.

2. Using the sum of your three skinfold site measurements, calculate your estimated body fat percentage using Table 11.1 or 11.2.

MALES			**FEMALES**	
Skinfold Sites	**Skinfold Thickness**		**Skinfold Sites**	**Skinfold Thickness**
abdominal	__________ mm.		triceps	__________ mm.
chest	__________ mm.		iliac crest	__________ mm.
thigh	__________ mm.		thigh	__________ mm.
total	__________ mm.		total	__________ mm.
% fat	__________		% fat	__________

3. Calculate your pounds of fat weight. Calculate your pounds of fat free weight.

   Fat Weight _______ lbs

   Fat Free Weight _______ lbs

4. Classify your body composition using the Health Fitness Standards in Table 7.2

5. Discuss your body composition relative to fitness standards. How does your body composition relate to your own health? How does it affect physical performance? If your fitness standard was below the good level for health, why?

6. Following the principles essential for good body composition;

   a. What aerobic exercises will you do (see Chapter 4)?

   b. What muscle development exercises will you do (see Laboratory 8)?

   c. What specific changes will you make in your diet?

TABLE 11.1   Percent Fat Estimates for Women Calculated from Triceps, Iliac Crest, and Thigh Skinfold Thickness

Sum of 3 Skinfold	<22	23-27	28-32	33-37	38-42	43-47	48-52	53-57	Over 58
23-25	9.7	9.9	10.2	10.4	10.7	10.9	11.2	11.4	11.7
26-28	11.0	11.2	11.5	11.7	12.0	12.3	12.5	12.7	13.0
29-31	12.3	12.5	12.8	13.0	13.3	13.5	13.8	14.0	14.3
32-34	13.6	13.8	14.0	14.3	14.5	14.8	15.0	15.3	15.5
35-37	14.8	15.0	15.3	15.5	15.8	16.0	16.3	16.5	16.8
38-40	16.0	16.3	16.5	16.7	17.0	17.2	17.5	17.7	18.0
41-43	17.2	17.4	17.7	17.9	18.2	18.4	18.7	18.9	19.2
44-46	18.3	18.6	18.8	19.1	19.3	19.6	19.8	20.1	20.3
47-49	19.5	19.7	20.0	20.2	20.5	20.7	21.0	21.2	21.5
50-52	20.6	20.8	21.1	21.3	21.6	21.8	22.1	22.3	22.6
53-55	21.7	21.9	22.1	22.4	22.6	22.9	23.1	23.4	23.6
56-58	22.7	23.0	23.2	23.4	23.7	23.9	24.2	24.4	24.7
59-61	23.7	24.0	24.2	24.5	24.7	25.0	25.2	25.5	25.7
62-64	24.7	25.0	25.2	25.5	25.7	26.0	26.2	26.4	26.7
65-67	25.7	25.9	26.2	26.4	26.7	26.9	27.2	27.4	27.7
68-70	26.6	26.9	27.1	27.4	27.6	27.9	28.1	28.4	28.6
71-73	27.5	27.8	28.0	28.3	28.5	28.8	29.0	29.3	29.5
74-76	28.4	28.7	28.9	29.2	29.4	29.7	29.9	30.2	30.4
77-79	29.3	29.5	29.8	30.0	30.3	30.5	30.8	31.0	31.3
80-82	30.1	30.4	30.6	30.9	31.1	31.4	31.6	31.9	32.1
83-85	30.9	31.2	31.4	31.7	31.9	32.2	32.4	32.7	32.9
86-88	31.7	32.0	32.2	32.5	32.7	32.9	33.2	33.4	33.7
89-91	32.5	32.7	33.0	33.2	33.5	33.7	33.9	34.2	34.4
92-94	33.2	33.4	33.7	33.9	34.2	34.4	34.7	34.9	35.2
95-97	33.9	34.1	34.4	34.6	34.9	35.1	35.4	35.6	35.9
98-100	34.6	34.8	35.1	35.3	35.5	35.8	36.0	36.3	36.5
101-103	35.2	35.4	35.7	35.9	36.2	36.4	36.7	36.9	37.2
104-106	35.8	36.1	36.3	36.6	36.8	37.1	37.3	37.5	37.8
107-109	36.4	36.7	36.9	37.1	37.4	37.6	37.9	38.1	38.4
110-112	37.0	37.2	37.5	37.7	38.0	38.2	38.5	38.7	38.9
113-115	37.5	37.8	38.0	38.2	38.5	38.7	39.0	39.2	39.5
116-118	38.0	38.3	38.5	38.8	39.0	39.3	39.5	39.7	40.0
119-121	38.5	38.7	39.0	39.2	39.5	39.7	40.0	40.2	40.5
122-124	39.0	39.2	39.4	39.7	39.9	40.2	40.4	40.7	40.9
125-127	39.4	39.6	39.9	40.1	40.4	40.6	40.9	41.1	41.4
128-130	39.8	40.0	40.3	40.5	40.8	41.0	41.3	41.5	41.8

Source: Jackson, Pollock & Ward. Generalized equations for predicting body density of women. Medicine and Science in Sports and Exercise, 12 (3):175-182, 1980.

TABLE 11.2   Percent Fat Estimates for Men Calculated from Chest, Abdomen, and Thigh Skinfold Thickness

Sum of 3 Skinfold	<19	20-22	23-25	26-28	Age Range 29-31	32-34	35-37	38-40
14-16	2.9	3.3	3.6	3.9	4.3	4.6	5.0	5.3
17-19	3.9	4.2	4.6	4.9	5.3	5.6	6.0	6.3
20-22	4.8	5.2	5.5	5.9	6.2	6.6	6.9	7.3
23-25	5.8	6.2	6.5	6.8	7.2	7.5	7.9	8.2
26-28	6.8	7.1	7.5	7.8	8.1	8.5	8.8	9.2
29-31	7.7	8.0	8.4	8.7	9.1	9.4	9.8	10.1
32-34	8.6	9.0	9.3	9.7	10.0	10.4	10.7	11.1
35-37	9.5	9.9	10.2	10.6	10.9	11.3	11.6	12.0
38-40	10.5	10.8	11.2	11.5	11.8	12.2	12.5	12.9
41-43	11.4	11.7	12.1	12.4	12.7	13.1	13.4	13.8
44-46	12.2	12.6	12.9	13.3	13.6	14.0	14.8	14.7
47-49	13.1	13.5	13.8	14.2	14.5	14.9	15.2	15.5
50-52	14.0	14.3	14.7	15.0	15.4	15.7	16.0	16.4
53-55	14.8	15.2	15.5	15.9	16.2	16.6	16.9	17.3
56-58	15.7	16.0	16.4	16.7	17.1	17.4	17.8	18.1
59-61	16.5	16.9	17.2	17.6	17.9	18.3	18.6	19.0
62-64	17.4	17.7	18.1	18.4	18.8	19.1	19.4	19.8
65-67	18.2	18.5	18.9	19.2	19.6	19.9	20.3	20.6
68-70	19.0	19.3	19.7	20.0	20.4	20.7	21.1	21.4
71-73	19.8	20.1	20.5	20.8	21.2	21.5	21.9	22.2
74-76	20.6	20.9	21.3	21.6	22.0	22.2	22.7	23.0
77-79	21.4	21.7	22.1	22.4	22.8	23.1	23.4	23.8
80-82	22.1	22.5	22.8	23.2	23.5	23.9	24.2	24.6
83-85	22.9	23.2	23.6	23.9	24.3	24.6	25.0	25.3
86-88	23.6	24.0	24.3	24.7	25.0	25.4	25.7	26.1
89-91	24.4	24.7	25.1	25.4	25.8	26.1	26.5	26.8
92-94	25.1	25.5	25.8	26.2	26.5	26.9	27.2	27.5
95-97	25.8	26.2	26.5	26.9	27.2	27.6	27.9	28.3
98-100	26.6	26.9	27.3	27.6	27.9	28.3	28.6	29.0
101-103	27.3	27.6	28.0	28.3	28.6	29.0	29.3	29.7
104-106	27.9	28.3	28.6	29.0	29.3	29.7	30.0	30.4
107-109	28.6	29.0	29.3	29.7	30.0	30.4	30.7	31.1
110-112	29.3	29.6	30.0	30.3	30.7	31.0	31.4	31.7
113-115	30.0	30.3	30.7	31.0	31.3	31.7	32.0	32.4
116-118	30.6	31.0	31.3	31.6	32.0	32.3	32.7	33.0
119-121	31.3	31.6	32.0	32.3	32.6	33.0	33.3	33.7
122-124	31.9	32.2	32.6	32.9	33.3	33.6	34.0	34.3
125-127	32.5	32.9	33.2	33.5	33.9	34.2	34.6	34.9
128-130	33.1	33.5	33.8	34.2	34.5	34.9	35.2	35.5

Source: Jackson and Pollock. Generalized equations for predicting body density of men. British Journal of Nutrition, 40:497-504, 1978.

Evaluating Cardiorespiratory Fitness

Name _______________________________ Section ____________ Date ____________

Introduction

Understanding one's cardiorespiratory fitness level is important in developing an exercise program. It is the foundation of determining the beginning intensity of an aerobic exercise program as well as charting progression throughout the phases of conditioning. It also allows the students to rate their level of aerobic capacity in relation to acceptable fitness standards.

Purpose

1. To help the students evaluate and rate their own level of cardiorespiratory fitness.

2. To understand the relationship between maximal oxygen uptake (consumption) and cardiorespiratory fitness.

3. To develop an awareness of satisfactory standards for cardiorespiratory fitness.

Procedures

1. Warm-up and stretch properly.

2. Accurately time a 1.5 mile Run/Walk at the fastest possible pace for this distance on a track or other precisely marked course.

3. Record time to the nearest second.

4. Cool down properly.

5. Using Tables 11.3 and 11.4 provided, determine one's level of fitness and the correlating estimated maximal oxygen consumption.

Note: This test should not be taken by anyone who has any health or medical problem that may be worsened by taking this test. Student should be "medically cleared" (Laboratory 1) before taking this test.

TABLE 11.3
1.5-MILE TIME AND MAXIMAL OXYGEN UPTAKE (VO₂ MAX ml/kg/min)

1.5-Mile Time min	VO$_2$ max	1.5-Mile Time min	VO$_2$ max
<7:30	>72	13:00	38
7:30	72	13:30	36
8:00	67	14:00	34
8:30	62	14:30	32
9:00	58	15:00	31
9:30	55	15:30	30
10:00	52	16:00	29
10:30	49	16:30	28
11:00	46	17:00	27
11:30	44	17:30	26
12:00	42	18:00	25
12:30	40	>18:00	<25

Adapted from Wilmore, Jack and Costill, David. <u>Training for Sport and Activity</u>. Third Edition. Dubuque: W. C. Brown. 1988. p. 368

TABLE 11.4

CARDIORESPIRATORY FITNESS STANDARDS ACCORDING TO
MAXIMAL OXYGEN UPTAKE (ml/kg/min)

FITNESS CATEGORY	AGE 18 - 29 Men	18 - 29 Women	30 - 39 Men	30 - 39 Women	40 - 49 Men	40 - 49 Women	50 - 59 Men	50 - 59 Women	60 - 69 Men	60 - 69 Women
Excellent	60	53	56	50	53	47	50	44	48	42
Very Good	53	46	49	43	45	39	43	37	41	35
Good	43	37	40	34	39	33	34	28	31	26
Fair	36	30	34	28	33	27	30	25	26	21
Poor	31	26	30	24	29	23	26	21	21	17
Very Poor	<31	<26	<30	<24	<29	<23	<26	<21	<21	<17

Adapted from Cooper, Ken. <u>The Aerobics Way</u>. New York: M. Evans & Company, 1977, pp. 280-281, and Hoeger, Werner and Hoeger, Sharon. <u>Fitness and Wellness</u>. Englewood: Morton Publishing Company, 1990, p. 15.

Results and Discussion

1. Determine your level of cardiorespiratory fitness and oxygen consumption. Use the appropriate tables (11.3 and 11.4).

   1.5 mile run time _________________ min:sec

   Maximal Oxygen Consumption _________________ ml/kg/min

   Cardiorespiratory Fitness level _________________

2. What is the relationship between maximal oxygen consumption and cardiorespiratory fitness?

3. Discuss your cardiorespiratory fitness relative to fitness standards. How does your cardiorespiratory fitness relate to your own health? If your fitness standard was below the good level for health, why?

Evaluating Muscular Strength/Endurance

Name _______________________________ Section _____________ Date _____________

Introduction

Developing muscular strength/endurance can enhance an individual's ability to resist fatigue and the ability to perform physical movement for extended periods of time without getting tired. In the absence of fatigue, there is greater success and enjoyment in daily work and recreational activities. Increasing one's strength can aid in the prevention of injuries, low back pain, poor posture, and other hypokinetic diseases as well as improve physical performance. In this lab, the students will be able to see how their level of muscular strength/endurance compares to fitness standards.

Purpose

1. To evaluate the student's muscular strength/endurance.

2. To understand the importance of muscular strength/endurance in an exercise program.

Procedures

1. Warm-up and stretch properly.

2. Students will find a partner to help with the test.

Sit-Up Test

1. Perform as many bent-leg sit-ups as possible within 60 seconds with your arms crossed in front of your chest touching your elbows to your thighs.

2. Appointed testers will time each test.

3. Partners will repeat the same procedure.

Push-Up Test

1. Support the body in a push-up position from the toes (men) or knees (women).

2. Keeping the body rigid, lower the body until the chest touches the floor. Repeat as many times as possible in 60 seconds with no more than 2 seconds rest between each push-up.

Pull-Up Test

1. Gripping a horizontal chinning bar with an over grip, pull your body up until the chin is above the bar, then lower yourself until the arms are fully extended. Repeat as many times as possible.

2. Females perform a flexed arm hang. Gripping a horizontal chinning bar with an over grip. Time begins when chin is placed over the bar with the arms flexed and stops when the chin drops below the bar.

Results and Discussion

1. Record the number of sit-ups, pull-ups, and push-ups completed and your fitness level for muscular strength/endurance (Refer to Table 11.5).

    Number of sit-ups ________________   Fitness Level ________________

    Number of pull-ups ________________   Fitness Level ________________

    Number of push-ups ________________   Fitness Level ________________

2. What is the relationship between muscular strength and endurance?

3. Discuss your muscular strength/endurance fitness relative to fitness standards. How does your muscular strength/endurance fitness relate to your own health?

4. If your fitness standard was below the good level for health, why?

Table 11.5 Fitness Standards For Sit-ups, Push-ups, and Pull-ups/Flexed Arm Hang Tests.

SIT-UPS STANDARDS

FITNESS LEVEL	15 - 19 Years		20 - 29 Years		30 - 39 Years		40 - 49 Years		50 - 59 Years		60 - 69 Years	
	Male	Female	Male	Female	Male	Female	Male	Female	Male	Female	Male	Female
Excellent	≥48	≥42	≥43	≥36	≥36	≥29	≥31	≥25	≥26	≥19	≥23	≥16
Good	42-47	36-41	37-42	31-35	31-35	24-28	26-30	20-24	22-25	12-18	17-22	12-15
Average	38-41	32-35	33-36	25-30	27-30	20-23	22-25	15-19	18-21	5-11	12-16	4-11
Fair	33-37	27-31	29-32	21-24	22-26	15-19	17-21	7-14	13-17	3-4	7-11	2-3
Poor	≤32	≤26	≤28	≤20	≤21	≤14	≤16	≤6	≤12	≤2	≤6	≤1

PUSH-UPS STANDARDS

FITNESS LEVEL	15 - 19 Years		20 - 29 Years		30 - 39 Years		40 - 49 Years		50 - 59 Years		60 - 69 Years	
	Male	Female	Male	Female	Male	Female	Male	Female	Male	Female	Male	Female
Excellent	≥39	≥33	≥36	≥30	≥30	≥27	≥25	≥22	≥21	≥20	≥18	≥16
Good	29-38	25-32	29-35	21-29	22-29	20-26	17-23	15-21	13-20	11-19	11-17	12-15
Average	23-28	18-24	22-28	15-20	17-21	13-19	13-16	11-14	10-12	7-10	8-10	4-11
Fair	18-22	12-17	17-21	10-14	12-16	8-12	10-12	5-10	7-9	2-6	5-7	2-3
Poor	≤17	≤11	≤16	≤9	≤11	≤7	≤9	≤4	≤6	≤1	≤4	≤1

PULL-UPS / FLEX ARM HANG STANDARDS

FITNESS LEVEL	15 - 19 Years		20 - 29 Years		30 - 39 Years		40 - 49 Years		50 - 59 Years		60 - 69 Years	
	Male	Female	Male	Female	Male	Female	Male	Female	Male	Female	Male	Female
Excellent	≥15	≥34	≥14	≥28	≥12	≥22	≥9	≥18	≥7	≥16	≥5	≥4
Good	12-14	20-33	10-13	18-27	8-11	15-21	5-8	11-17	4-6	8-15	3-4	6 13
Average	8-11	10-19	6-9	8-17	4-8	7-14	2-4	4-10	2-3	1-7	1-2	1-5
Fair	5-7	3-9	2-5	2-7	1-3	1-6	1	1-3	1	0	0	0
Poor	0-4	0-2	0-2	0-1	0	0	0	0	0	0	0	0

Evaluating Flexibility

Name _________________________________ Section _____________ Date _____________

Introduction

Flexibility has traditionally been the most neglected of the health-related components of physical fitness. However, it is recognized as an integral part of any proper exercise program. Increased range of motion is thought to aid in the prevention of injuries associated with loss of mobility. The importance of this lab is to develop an awareness of one's level of flexibility.

Purpose

1. To evaluate and rate the student's level of flexibility.

2. To understand what is an acceptable range of motion.

Procedure

1. Warm-up and stretch properly.

2. Students will find a partner to help with the three flexibility tests.

3. Follow the directions below to complete all three tests.

SIT-AND-REACH TEST
DIRECTIONS: The sit-and-reach test measures the flexibility of the lower back and hamstrings.
a.   Sit on the floor with shoes removed and facing the flexibility box.
b.   With knees fully extended and feet four inches apart, place the box against the flat bottom of the feet.
c.   Extend arms straight forward with one hand placed on top of the other, palms down.
d.   Reach forward as far as possible and hold the position until measured to the nearest half inch.
e.   The best of three trials is recorded. The partner makes sure both hands remain even and the legs remain straight by placing a hand on the knees.

SHOULDER LIFT TEST
DIRECTIONS: The shoulder lift test measures the flexibility of the shoulders.
a.   Lie on your stomach with arms extended and holding a straight edge.
b.   With the forehead maintaining contact with the floor, raise the arms as high as possible.
c.   The greatest distance achieved between the floor and the straight edge is recorded of the three trials.

ACHILLES STRETCH TEST

DIRECTIONS: The Achilles stretch test measures the flexibility of the ankles.

a.   Stand facing the wall with extended arms and legs straight with toes facing forward.

b.   Lean against the wall with the hands and press the hips to the wall while maintaining straight legs.

c.   Your partner will measure the ankle angle with a goniometer and record the best of three trials.

TABLE 11.6

Flexibility Standards for Sit-and-Reach, Shoulder Lift, and Achilles Stretch

FLEXIBILITY STANDARDS

Fitness Level	Sit-and-Reach	Shoulder Lift	Achilles Stretch
Excellent	>23 inches	>26 inches	<60 degrees
Very Good	21.5 - 23 inches	24.5 - 26 inches	60 - 69 degrees
Good	18.5 - 21 inches	21.5 - 24 inches	70 - 79 degrees
Average	16.5 - 18 inches	18.5 - 21 inches	80 - 89 degrees
Fair	13.5 - 16 inches	14.5 - 18 inches	90 - 99 degrees
Poor	11.5 - 13 inches	10.5 - 14 inches	100 - 110 degrees
Very Poor	<11 inches	<10 inches	>110 degrees

Results and Discussion:

1.   Record your results below from the flexibility tests and determine your level of flexibility.

Sit-and-Reach Score____________   Fitness Level__________________

Shoulder Lift Score____________   Fitness Level__________________

Achilles Stretch Score__________   Fitness Level__________________

2.   Discuss your flexibility relative to fitness level. How does your flexibility relate to your own health?

3.   If your fitness level was below the good category for health, why?

Measuring One's Heart Rate

Name ________________________________ Section ____________ Date ____________

Introduction

The amount of oxygen consumed by the body is proportionate to heart rate. For this reason, heart rate gives a simple and readily available index of cardiac stress.

Purpose

1. Become accurate in detecting and counting heart rate.

2. To demonstrate the effect of participating in various activities on heart rate.

Procedure

1. Students will find their heart rate after each activity and count for 10 seconds and multiply by 6 for calculating heart rate for one minute.

2. Students record the results and answer the questions.

Results and Discussion

1. Measure your resting heart rate. To get a valid resting measure, record your heart rate five days in a row in the morning before you get out of bed.

   Record the lowest value.

   What health fitness category is it (see Table 3.1)?

2. Measure your heart rate response to physical activity.

   a. Standing at attention   ____________________

   b. Casual walk (3-4 min.)   ____________________

   c. Jog in place (3 min.)   ____________________

3. Compare your heart rate at rest to the various forms of physical activity. Did your heart rate increase or decrease? Why this change?

4. Would persons sometimes feel faint if they were jogging hard and suddenly stopped and stood still? Why?

Prescribing an Aerobic Exercise Program

Name _________________________________ Section ____________ Date ____________

Introduction

Most beginners either exercise too intensely and have to stop before they receive a sufficient benefit from exercise, or they do not exercise at a sufficient overload to obtain cardiorespiratory improvement. Students should target their exercise heart rate to be between 70 and 85 % of their maximum exercise heart rate. The emphasis on an aerobic exercise program should be on extending the duration time of the exercise without tiring.

Purpose

1. To practice taking exercise heart rate.

2. To learn what happens to your heart rate when walking/jogging at different paces.

3. To determine the walking/jogging pace that will produce an exercise heart rate that will give you the training effect.

4. To figure your exercise target heart rate.

5. To select an appropriate progression rate for an aerobic exercise program.

Procedure

1. Calculate your predicted maximal heart rate and your 50, 60, 70, and 80, and 90 % heart rate intensities. (Results 1 and 2 below.)

2. Do general warm-up exercise for five minutes.

3. Walk/Jog for about 3 minutes at a specific per-mile pace. (16, 14, 12, 10, 8, and 6 minutes/mile pace.)

4. Immediately after completing the distance, take heart rate for 10 seconds and multiply by six to get exercise heart rate per minute.

5. Plot heart rate for each pace on the graph.

6. Rest 2-3 minutes before going at next faster pace.

7. From graph, determine pace at which to walk/jog in order to be exercising at 50, 60, 70, 80, and 90 % of your maximum.

8. Prescribe a personal aerobic exercise program.

Results and Discussion

1. Calculate your Maximal Heart Rate (MHR)

   220 – age = _____________________ MHR

2. Calculate your Exercise Heart Rate at the following intensities:

   MHR x .50 = _____________________ 50% Intensity

   MHR x .60 = _____________________ 60% Intensity

   MHR x .70 = _____________________ 70% Intensity

   MHR x .80 = _____________________ 80% Intensity

   MHR x .90 = _____________________ 90% Intensity

3. Plot heart rate for the following paces:

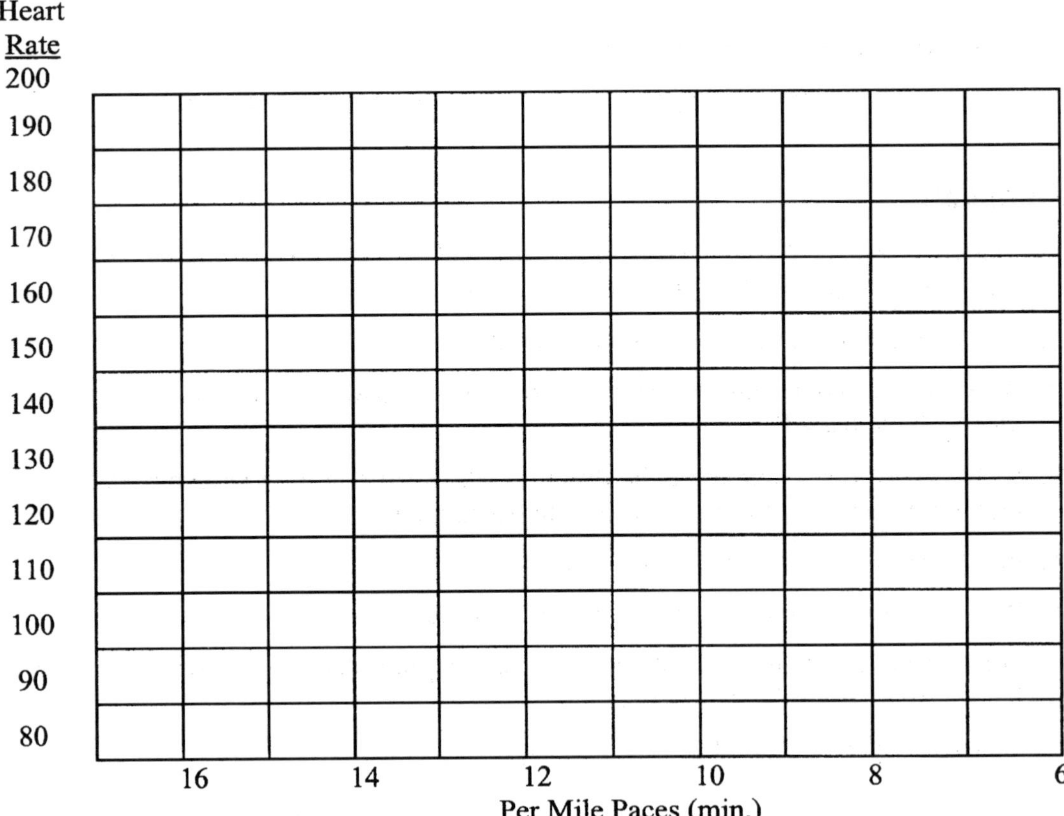

4. What pace do you need to walk/jog to be at a selected exercise training heart rate and what is your oxygen consumption in ml/kg/min at that pace? (For oxygen uptake values, estimate on the basis of your maximal oxygen uptake calculated from the 1.5 mile field test.)

Heart Rate Intensity	Per Mile Pace	Oxygen uptake (ml/kg/min)
Maximum Heart Rate		
85%		
80		
75		
70		
65		
60		
55		

5. What training heart rate level do you feel is appropriate for you as you begin your exercise program?

   Training Heart Rate ___________________   Why?

6. Following the principles of conditioning (overload, specificity, and individuality), prescribe yourself an aerobic exercise program to improve or maintain your cardiorespiratory fitness. Your goal for the future should be to exercise by walking, jogging, swimming, cycling, or a comparable aerobic activity three to five days per week for 60 minutes at 60-65 %, or 45 minutes at 70-75%, or 30 minutes at 80-85%.

   What days of the week will you exercise? ___________________

   What time of the day will you exercise? ___________________

   Select the appropriate starting level (from Table 6.1 and Figures 6.10 and 6.11).

   Starting level: ___________________

   Rationale for selecting the starting level:

7. Considering your need, interests, and the availability of the activity, what will be the best aerobic exercise for you to do and why?

8. Record your aerobic activities over a six week period using the log located in Laboratory 13.

Prescribing a Muscular Strength/Endurance Exercise Program

Name _________________________ Section ___________ Date ___________

Introduction

Weight training is probably the best exercise method for increasing strength/muscular endurance. Other benefits of weight training include the prevention of injury, the strengthening of bones, and aid in controlling body composition. This lab provides some basic resistance training exercises that can be implemented into an total exercise program.

Purpose

1. To provide the student instruction in safe techniques for weight training.

2. To learn basic resistance training exercises for the major muscle groups.

3. To prescribe an exercise program for the improvement or maintenance of muscular strength/endurance.

Procedures

1. Warm-up and stretch properly.

2. Instructor will demonstrate correct techniques in selected weight training exercises.

3. Students will experiment and perform each exercise.

4. Prescribe a personal muscular strength/endurance exercise program.

Results and Discussion

1. When developing a weight training program for health fitness, all major muscle groups should be overloaded. To determine the appropriate resistance for each set, find your 10 RM weight for each exercise below. Then use 75 % of the 10 RM weight for the first set followed by the 10 RM weight for the second set.

Exercise	Weight for set 1	Weight for set 2
Bench Press	__________	__________
Lat Pull	__________	__________
Elbow extensions	__________	__________
Arm Curls	__________	__________
Knee extensions	__________	__________
Knee Curls	__________	__________

2. Following the principles of conditioning (overload, specificity, and individuality), prescribe yourself a weight training program to improve or maintain your muscular strength/endurance fitness.

What days of the week will you exercise: ___________________________________

What time of the day: ___________________________________

3. Which area of the body do you feel you have below average strength? Why?

4. Why is it important to develop or have good muscular strength/endurance?

5. Record your muscular strength/endurance activities over a six week period using the log in Laboratory 13.

Prescribing a Flexibility and Back Exercise Program

Name _______________________________ Section ___________ Date ___________

Introduction

Flexibility is an integral part of an exercise program. Increasing one's range of motion will allow joints to move with greater ease and efficiency, as well as decreasing the chance of injuries. During the warm-up light stretching helps to prepare the body for the work-out activity. When cooling down after the work-out flexibility training should be performed. The importance of this laboratory is to provide an opportunity to experience and distinguish between the different types of flexibility exercises and present a safe flexibility program that can be continued throughout ones lifetime.

Included in this laboratory are also several back exercises. At some point in their lives, approximately 80 percent of all adults will experience back pain that limits their ability to function normally. An estimated 75 million Americans have recurring back problems, and two million cannot hold jobs as a result of low back pain causing 93 million days of lost work per year and $10 billion in worker's compensation. Proper care of the back is essential to good health and the addition of back exercises to an exercise program can be a preventive measure to one's future health.

Purpose

1. To provide the student with safe range of motion exercises that can be incorporated into their fitness program.

2. To learn the importance of and understand techniques for a warm-up and cool-down in an exercise program.

3. To provide the student with safe exercises for preventing or alleviating low back pain.

4. To learn proper methods for strengthening the lower back.

5. To learn the importance of proper lower back care.

Procedures

1. Students will walk or jog slowly for three to five minutes as a general warm up.

2. Instructor will provide instruction and explanation for proper utilization of the below flexibility and back exercises.

Flexibility Exercises (see Chapter 6)   **Back Exercises (see Chapter 10)**

Flexibility Exercises (see Chapter 6)	Back Exercises (see Chapter 10)
Shoulder stretch across	Curl-up
Shoulder stretch behind	Pelvic raises
Shoulder stretch side	Side leg raises
Trunk stretch to side	Knees to chest
Hip & trunk stretch	Back extensions
Hamstring stretch	
Groin stretch	
Thighs stretch (quadriceps)	
Achelles stretch (Calf)	

Results and Discussion

1. Following the principles of conditioning (overload, specificity, and Individuality), prescribe yourself a program to improve or maintain your flexibility and strength of your back.

   When will you perform your stretching and back exercises?

   Days/week ________________________________ , Time of day ____________________

2. Why should you perform flexibility and back exercises weekly?

3. Have you ever had low back pain? If so, how did you get rid of it?

4. List the flexibility and back exercises you will do over a six week period. Using the log in Laboratory 13 record the days you actually do the exercises and list the repetitions.

Estimating Caloric Expenditure

Name _________________________________ Section ____________ Date ____________

Introduction

Controlling one's weight and body fat can be accomplished best through diet and exercise. To minimize loss of fat free weight, a weight reduction of no more than 1-2 pounds a week is recommended. Although the elimination of extra calories in the amount food consumed per day can be achieved quiet easily, the caloric expenditure during exercise is more difficult. However, weight lost through aerobic activities and strength training will insure a greater loss of fat and a more permanent fat loss. By monitoring intensity and duration of exercise, caloric expenditure can be estimated. A negative caloric balance of 3500 calories is required to lose one pound of fat.

Purpose

1. To learn how to calculate the estimated caloric expenditure of fitness activities.

2. To understand the relationship between intensity and caloric expenditure.

3. To determine the amount of exercise per week needed to lose a pound of fat.

Procedures

1. Calculate your Resting Metabolic Rate (RMR) using your fat-free body (FFM) weight from Laboratory 3 and Table 11.7.

2. Determine your daily physical exertion level (DPEL) using Table 11.8.

3. Calculate the caloric expenditure of your fitness activities using Table 11.9.

4. Using the equation, determine your daily caloric expenditure.

Results and Discussion

1. Calculate your RMR from Table 11.7 (divide LBS by 2.2 for Kg).   ________

   Multiply by your DPEL activity factor (Table 11.8).   X ________

   This equals your daily caloric expenditure without exercise (DCEE)   ____

TABLE 11.7   Estimation of Resting Metabolic Rate (RMR) in Calories Based on Fat-Free Body Mass (FFM) in Kilograms.

FFM(Kg)	RMR(Cal)	FFM(Kg)	RMR(Cal)	FFM(Kg)	RMR(Cal)	FFM(Kg)	RMR(Cal)
40	1234	55	1558	70	1882	85	2206
41	1256	56	1580	71	1904	86	2228
42	1277	57	1601	72	1925	87	2249
43	1299	58	1623	73	1947	88	2271
44	1320	59	1644	74	1968	89	2292
45	1342	60	1666	75	1990	90	2314
46	1364	61	1688	76	2012	91	2336
47	1385	62	1709	77	2033	92	2357
48	1407	63	1731	78	2055	93	2379
49	1428	64	1752	79	2076	94	2400
50	1450	65	1774	80	2098	95	2422
51	1472	66	1796	81	2120	96	2444
52	1493	67	1817	82	2141	97	2465
53	1515	68	1839	83	2163	98	2487
54	1536	69	1860	84	2184	99	2508

TABLE 11.8   Classification of Daily Physical Exertion Level (DPEL) Including Work but Excluding Exercise.

DPEL Classification	Activity Factor	Most of Your Day Consists of
Sedentary	1.3	Office/Computer Work, Watching TV, Driving, Studying, Reading, Cooking, and etc.
Moderate	1.5	House Cleaning, Walking, Yard Work, Climbing Stairs, Building Trades Occupation, and etc.
Active	1.7	Military on Active Duty, Manual Laborer, Lumberjack, and etc.

2. Calculate the number of calories expended (CE) per fitness activity (Table 11.9).

Fitness Activity	Cal/min/Kg	Body Wt (Kg)	Time (min)	CE
_______	_______	X _______	X _______	= _______
_______	_______	X _______	X _______	= _______
_______	_______	X _______	X _______	= _______
_______	_______	X _______	X _______	= _______

Fitness Activities Total Caloric Expenditure (FATCE)   _______

**TABLE 11.9 Common Fitness Activities and Caloric Expenditure Per Minute
Body Weight in Kilograms**

Fitness Activities	Cal/min/Kg	Fitness Activities	Cal/min/Kg
Basketball		Music Aerobics	
½ court	.13	Moderate	.10
Full court	.15	Intense	.13
Cycling		Racquetball	
Leisure < 10 mph	.06	Singles	.15
Intense 10-20 mph	.11	Doubles	.10
Racing > 20 mph	.17	Resistance Training	.08
Golf		Swimming (Fitness)	.15
Walking	.08	Tennis	
Riding	.04	Singles	.14
Jogging (on level)		Doubles	.10
6 min/mile	.25	Walking (on level)	
7 min/mile	.22	14 min/mile	.11
8 min/mile	.21	15 min/mile	.095
9 min/mile	.19	16 min/mile	.09
10 min/mile	.17	17 min/mile	.085
11 min/mile	.15	18 min/mile	.08
12 min/mile	.14	19 min/mile	.075
13 min/mile	.12	20 min/mile	.07

3. Add to DCEE the total caloric expenditure of all fitness activities and Calories
   expended during Thermogenesis (200 for males and 150 for females) to determine
   the daily total caloric expenditure.

    DCEE   ____________

    + FATCE   ____________

    + Thermo   200 or 150

    Total Cal   ____________

4. How is exercise intensity related to caloric expenditure?

5. Which type of fitness activities use the most Calories per minute?

6. How does muscle development relate to basal metabolic rate and resting metabolic rate?

7. How does fitness activities affect your caloric expenditure while at rest?

Three-Day Nutritional Analysis

Name _________________________________ Section ____________ Date ____________

1. On the basis of the New USDA Food groups, record your intake of all food and fluids for three days. List the food and the number of servings you had for each category.

FOOD GROUP	DAY 1		DAY 2		DAY 3	
	Food	#Servings	Food	#Servings	Food	#Servings
Breads, cereals and other grain products: Whole grain or enriched						
Fruit: Citrus, melons, berrics, all other fruit						
Vegetables: Dark green leafy, deep yellow, all other vegetables						
Protein Group 1: Lean meat, poultry, fish, or Eggs Beans and peas						
Protein Group 2: Lowfat milk, cheese, yogurt						
Fats (unsaturated): Vegetable oils, salad dressing, spreads, nuts						
Sweets and alcoholic beverages:						

2. Using the "suggested daily servings" found below, list the number of actual servings you had from each food group and compare to the suggested daily serving (use an average from the three day nutritional analysis).

Food Group	**Suggested Daily Servings**	**Number of Actual Servings**
Bread and Cereals	6-11	________________
Fruit	2-4	________________
Vegetables	3-5	________________
Protein group 1	2-3	________________
Protein group 2	2-4	________________
Fats	limit	________________
Sweets and Alcohol	limit	________________

3. Determine your daily need for protein as follows:

Step 1.

Weight in pounds _______ ÷ by 2.2 = _______ weight in Kilograms

Step 2.

Weight in Kilograms _____ x (.8-1.2 normal adult) = ______ grams of protein needed each day

Weight in Kilograms _____ x (1.2-1.5 athlete) = ______ grams of protein needed each day

Weight in Kilograms _____ x (1.5-2.0 pregnant) = ______ grams of protein needed each day

4. On the basis of Table 9.6, what should be your daily intake of fat in grams? _______

5. Summarize your food intake from the chart for three days.

NUTRIENT	Average Grams/Day	Calories/gram	Average Calories/Day	Percent of Total Calories
Protein		times 4		
Fat		times 9		
Carbohydrates		times 4		
Fiber		NA	NA	NA
TOTAL	NA	NA		100

6. Analyze your diet. Are you meeting the recommendations in terms of grams of proteins, carbohydrates, fiber, and fat per day? Are you eating the suggested number of servings of the food groups as recommended in Chapter 9? What are the areas where you need to improve? What lifestyle changes do you need to make to have a healthy diet?

Synthesis of Fitness Goals, Exercise Program, and Results

Name _______________________________ Section ____________ Date ____________

I. Pre-Test Health-Fitness Results

Cardiorespiratory Fitness

1.5 mile run time

Maximal Oxygen Consumption ___________ ml/kg/min

Cardiorespiratory Fitness Level   ____________

Muscular Strength/Endurance Fitness

Number of sit-ups	____________	Fitness Level	____________
Number of pull-ups	____________	Fitness Level	____________
Number of push-ups	____________	Fitness Level	____________

Flexibility Fitness

Sit-and-Reach Score	____________	Fitness Level	____________
Shoulder Lift	____________	Fitness Level	____________
Achelles Stretch	____________	Fitness Level	____________

Body Composition Fitness

% fat ____________   Fitness Level   ____________

II. Objectives

Overall Goals (circle one)

Cardiorespiratory Fitness	Improve	Maintain
Muscular Strength/Endurance Fitness	Improve	Maintain
Flexibility Fitness	Improve	Maintain
Body Composition Fitness	Improve	Maintain

Specific Objectives

Over the next six weeks, what are your specific objectives for the health-fitness tests that will be administered at the end of the course?

Cardiorespiratory Fitness

1.5 mile run time   ________________   Fitness Level

Muscular Strength/Endurance Fitness

Number of sit-ups   ____________   Fitness Level   ____________

Number of pull-ups   ____________   Fitness Level   ____________

Number of push-ups   ____________   Fitness Level   ____________

Flexibility Fitness

Sit-and-Reach Score   ____________   Fitness Level   ____________

Shoulder Lift   ____________   Fitness Level   ____________

Achelles Stretch   ____________   Fitness Level   ____________

Body Composition Fitness

% fat ____________   Fitness Level   ____________

III. Weekly Physical Activity Format

In order to improve in health fitness, a weekly physical activity schedule needs to be planned. If not planned into one's day, exercise often is neglected. How will you schedule your exercise program? On which days will you exercise? What health fitness components will you exercise on which days? Fill in the chart below with your weekly exercise plan.

Activity	Monday	Tuesday	Wednesday	Thursday	Friday	Saturday	Sunday
Aerobic							
MS/ME							
Flexibility							
Sports							

162

IV. Exercise Logs

AEROBIC ACTIVITY LOG

Week		Sun	Mon	Tues	Wed	Thurs	Fri	Sat
One	Activity							
	Intensity*							
	Time							
Two	Activity							
	Intensity							
	Time							
Three	Activity							
	Intensity							
	Time							
Four	Activity							
	Intensity							
	Time							
Five	Activity							
	Intensity							
	Time							
Six	Activity							
	Intensity							
	Time							

*Record average heart rate for the entire time of the activity.

MUSCULAR STRENGTH/ENDURANCE WEIGHT TRAINING LOG*

Week	One		Two		Three		Four		Five		Six	
	reps/wt		reps/wt		reps/wt		reps/wt		reps/wt		reps/wt	
Exercise	days/wk		days/wk		days/wk		days/wk		days/wk		days/wk	

*For reps/wt, record the number of repetitions you do on your last set and the 10 RM weight.

V. Post-Test Results

Cardiorespiratory Fitness

1.5 mile run time

Maximal Oxygen Consumption _____________ ml/kg/min

Cardiorespiratory Fitness Level _____________________

Muscular Strength/Endurance Fitness

Number of sit-ups _______________ Fitness Level _______________

Number of pull-ups _______________ Fitness Level _______________

Number of push-ups _______________ Fitness Level _______________

FLEXIBILITY LOG

Week	One		Two		Three		Four		Five		Six	
Exercise	days/wk		days/wk		days/wk		days/wk		days/wk		days/wk	

Flexibility Fitness

Sit-and-Reach Score ____________ Fitness Level ____________

Shoulder Lift ____________ Fitness Level ____________

Achelles Stretch ____________ Fitness Level ____________

Body Composition Fitness

% fat ____________ Fitness Level ____________

VI. Analysis of Results

On a separate sheet of paper, respond to the following questions in each area.

Overall

What types of exercises did you implement in your program? Explain your typical workout for each day and for the week. Were records or logs kept? Was there any variation?

Cardiorespiratory Component

Was there improvement? Did you meet your goal? What changes occurred in heart rate, blood pressure, oxygen consumption, fitness level as well as other physiological effects. How did your choice of activity affect these changes? How did you implement the principles of exercise in your program?

Muscular Strength/Endurance Component

Was there improvement? Did you meet your goal? What changes occurred in muscular strength/endurance, girth, weight, etc.? How did your choice of activity affect these changes? How did you implement the principles of exercise in your program?

Flexibility Component

Was there improvement? Did you meet your goal? What changes occurred in range of motion? How did your choice of activity affect these changes? How did you implement the principles of exercise in your program?

Body Composition Component

Was there improvement? Did you meet your goal? What changes occurred in weight and body measurements? How did your choice of activity affect these changes? How did you implement the principles of exercise in your program? Did you change your diet?

Summary and Conclusions

What did you attribute to the failure or success of your program? What changes need to be made to improve your current fitness level? How will you incorporate physical activity into your lifestyle in the future?

Bibliography

Allsen, P. E. *Strength Training—Beginners, Bodybuilders, and Athletes*. Glenview, IL: Scott, Foresman and Company, 1987.

Alter, M. J. *Science of Flexibility*. (2nd Ed.) Champaign, IL: Human Kinetics, 1996.

American Association for Active Lifestyles. *Physical Best and Individuals with Disabilities*. Reston, VA, 1995.

American College of Sports Medicine. *ACSM's Guidelines for Exericse Testing and Prescription (5th Ed.)*. Baltimore: Williams and Wilkins, 1995.

Anderson, B. *Stretching*. Bolina, CA: Shelter Publications, 1980.

Armstrong, R. B. "Mechanisms of exercise-induces delayed onset muscular soreness: a brief review." *Medicine and Science in Sports and Exercise*. 16(1984):529-538.

Astrand, P.O. and Rodahl, K. *Textbook of Work Physiology*. 3rd Ed. New York: McGraw-Hill, 1986.

Beaulieu, J. *Stretching for All Sports*. Pasadena, CA: The Athletic Press, 1980.

Benfante R, Reed D. "Is elevated serum cholesterol a risk factor for coronary heart disease in the elderly?" *Journal of the American Medical Association*. 263 (1990): 393-96.

Bensley, Robert J. "Defining Spiritual Health: A Review of the Literature." *Journal of Health Education* 22 (5) (1991): 287-290.

Berger, Bonnie G., Friedmann, Erika, Eaton, Muzza. "Comparison of Jogging, the Relaxation Response, and Group Interatction for Stress Reduction". *Journal of Sport & Exercise Psychology* 10 (1988): 431-447.

Bishop, J.G. & Aldana, S.G. *Step Up to Wellness: A Stage Based Approach*. Boston, MA: Allyn and Baoon, 1999.

Blair S.N , Kohl H.W., Paffenbarger R.S., et al. "Physical fitness and all-cause mortality." *Journal of the American Medical Association*. 262 (1989): 2395-2401.

Blankhenhorn D.H., Johnson R.L., Mack W.J., et al. "The Influence of Diet on the Appearance of New Lesions in Human Coronary Arteries." *Journal of American Medical Association*. 263 (1990): 1646-52.

Blazer D.G. "Social Support and Mortality in an Elderly Community Population." *American Journal of Epidemiology*. 115(5) 1982:684-94.

Blumer, D., and Heilbronn, M. "Chronic back pain as a variant of depressive disease: The pain prone disorder" *Journal of Nervous and Mental Disorders*. 170(1982):381.

Borg, G. A. Psychological Basis of Physical Exertion. *Medicine and Science in Sport and Exercise* (1982): 377.

Braith, R.W., Graves, J. E., Pollock, M.L., Leggett, S. L., Carpenter, D. M., and Calvin, A.B. "Comparison of two versus three days per week of variable resistance training during 10 and 18 wee programs." *International Journal of Sport Medicine*. 10(1989):450-454.

Brand, Paul, and Philip Benson H. *Beyond the Relaxation Response.* New York: Times Books, 1984.

Brodsky M.A., Sato D.A., Iseri L.T., et al. "Ventricular Tachyarrythmia Associated with Psychological Stress. The Role of the Sympathetic Nervous System." *Journal of the American Medical Association.* 257(15) (1987): 2064-67.

Brooks, G. A., and T. D. Fahey. *Exercise Physiology: Human Bioenergetics and Its Applications.* New York: John Wiley & Sons, 1984.

Byrd R.C. "Positive Therapeutic Effects of Intercessory Prayer in a Coronary Care Unit Population." *Southern Medical Journal.* 81(7) 1988:826-29.

Carver, S. "Injury Prevention and Treatment." In E. T. Howley & B. D. Franks, *Health/Fitness Instructor's Handbook.* Champaign, IL: Human Kinetics, 1986: 211-232.

Chodzko-Zajko, W.J. "The Physiology of Aging and Exercise", in R.T.Cotton & C. Ekeroth (Eds). *Exercise for Older Adults. Champaign,* IL: Human Kinetics, 1998.

Clark, Nancy. "Fueling Up With Carbs: How Much Is Enough? *The Physician and Sports Medicine* 19(1991)8: 68-69.

Cooper, K. H., et al. "Physical Fitness Levels and Selected Coronary Risk Factors. A Crossection Study." *Journal of the American Medical Association* 236 (1976): 66.

Cordain, Loren, et al. "The Effects of an Aerobic Running Program on Bowel Transit Time." *Journal of Sports Medicine* 26(1986):101-104.

Croft J.B., Cresanta J.L., Webber L.S., et al. "Cardiovascular Risk in Parents of Children with Extreme Lipoprotein Levels: the Bogalusa Heart Study." *Southern Medical Journal.* 81 (1988):341-49, 353.

Drenick, Ernst J., et al. "Excessive Mortality and Causes of Death in Morbidly Obese Men." *Journal of American Medical Association* (February 1, 1980): 443-45.

Dressendorfer, Rudolph H. "Physiological Profile of a Masters Runner." *The Physician and Sportsmedicine* 8 (August 1980): 49-52.

Dwyer, Terry, et al. "A Comparison of Trends of Coronary Heart Disease Mortality in Australia, USA, England, and Wales with Reference to Three Major Risk Factors--Hypertension, Cigarette Smoking, and Diet." *International Journal of Epidemiology 9* (1980): 65-71.

Dwyer J.T. "Health Aspects of Vegetarian Diets." *American Journal of Clinical Nutrition.* 1988;48(3 Suppl)712-38.

Edlin, G., Golanty, E., & Brown, K. *Health and Wellness.* (6th Ed.) Sudbury, MA: Jones and Barlett Publishers, 1999.

Enstrom, James E. "Cancer Mortality Among Mormons in California During 1968-1975." *Journal* of *Clinical Investigation* 65 (November 1980): 1073-82.

Epstein, L.H., et al. "Aerobic Exercise and Weight." *Addictive Behaviors* 5 (1980): 371-88.

Epstein, L.H., et al. "The Effects of Contract and Lottery Procedures on Attendance and Fitness in Aerobic Exercise." *Behavior Modification* 4 (1980): 465-80.

Epstein, L.H., et al. "A Comparison of Lifestyle Change and Programmed Aerobic Exercise on Weight and Fitness Changes in Obese Children" *Behavior Therapy* 13 (1982): 651-65.

Erikssen, Jan, et al. "Coronary Risk Factors and Physical Fitness in Healthy Middle-Aged Men" *Acta Medica Scandinavica* 645 (1981): 57-64.

Fleck, S.J. and Kreamer, W. J. *Designing Resistance Training Programs.* Champaign, IL: Human Kinetics Books, 1987.

Foss, M.L. & Keteyian, S.J. Fox's Physiological Basis for Exercise and Sport. (6th Ed.). Boston, MA: WCB McGraw-Hill, 1998.

Fox, E. L., R. W. Bowers, and M. L. Foss. *The Physiological Basis for Exercise and Sport.* Wisconsin: Brown & Benchmark, 1993.

Fox, E. L., T. E. Kirby, and A. R. Fox. *Bases of Fitness.* New York: Macmillan Publishing Company, 1987.

Gettman, L. R., Ward, P. and Hagman, R. D. "A comparison of combined running and weight training with circuit weight training." *Medicine and Science in Sports and Exercise.*14(1982): 229-234.

Gillam, G.M. "Effects of frequency of weight training on muscle strength enhancement." *Journal of Sports Medicine.* 21(1981): 432-436.

Graves, J. E., Pollock, M.L., Leggett, S. H., Braith, R.W., Carpenter, D. M., and Bishop, L. E. "Effect of reduced training frequency on muscular strength." *International Journal of Sports Medicine.* 9(1988):316-319.

Graves, J. E., Pollock, M.L., Jones, A. E., Calvin, A.B., and Legett, S. H. "Specificity of limited range of motion variable resistance training." *Medicine and Science in Sports and Exercise.* 21(1989):84-89.

Griffin G.C. and Catelli W.P. *How to Lower Your Cholesterol and Beat the Odds of a Heart Attack.* Tucson: Fisher Books, 1989.

Gyntelberg, F., et al. "Blood Pressure Reduction by Change in Life Style: The CVD Intervention Study in Glostrup." *Acta Medica Scandinavica* 646 (1980): 10-14.

Hafen, Brent Q. *Nutrition, Food, and Weight Control.* Newton, Mass.: Allyn & Bacon, 1981.

Hakkinen, K. "Factors influencing trainability of muscular strength during short term and prolonged training." *National Strength and Conditioning Association Journal.* 7(1985):32-34.

Hartung, Harley G. et al. "Relation of Diet to High-Density-Lipoprotein Cholesterol in Middle-aged Marathon Runners, Joggers, and Inactive Men." *The New England Journal of Medicine* 302 (1980): 357-61.

Haskell, W.L. "Physical Activity in the Prevention and Management of Coronary Heart Disease." *PCPFS Physical Activity and Fitness Research Digest, 2 (1) 1995.*

Heyward, V.H. *Adbvanced Fitness Assessment & Exercise Prescription. (3rd Ed.). Champaign, IL: Human Kinetics, 1998.*

Hoeger, W. W. K., & Hoeger, S.A.. *Fitness and Wellness.* Englewood, CO: Morton Publishing Company, 1990.

Hoeger, W.W.K. & Hoeger, S.A. *Principles & Labs for Fitness & Wellness. (5th Ed.).* Englewood, CO: Morton Publishing Company, 1999.

Houtkooper, Linda. "Food Selection for Endurance Sports." *Medicine and Science In sports and Exercise* 24(1992) S349-S355.

Howley, E. T. and Franks B.D. *Health Fitness Instructor's Handbook, Champaign, IL.* Human Kinetics. 1997.

Huber, F & Teohanchuk, P. *Essentials of Physical Activity: Laboratory Manual.* (3rd Ed.) Dubuque, IA: Eddie Bowers Publishing, 1999.

Hubert, Helen B. "Obesity as an Independent Risk Factor for Cardiovascular Disease: A 26-Year Follow-up of Participants in the Framingham Heart Study." *Circulation* 67 (May 1983): 968-76.

International Society of Sport Psychology. "Physical Activity and Psychological Benefies." *The Physician and Sportsmedicine.* 20 (October, 1992): 179-184.

Jasnoski M.L., Kugler J. "Relaxation, Imagery, and Neuroimmunomodulation." *Annals of the New York Academy of Science.* 496 (1987): 722-30.

Jones, D.A., Newman, D.J., Round, J. M., and Tolfree, S. E. L. "Experimental human muscle damage: morphological changes in relation to other indices of damage." *Journal of Physiology.* 375(1986):435-438.

Kanvonen, Karen, et al. "Myth Busters." *Women's Sports and Fitness* October 1991.

Kaplan G.A., Salonen J.T., Cohen R.D., Brand R.J., et al. "Social Connections and Mortality From All Causes and From Cardiovascular Disease: Prospective Evidence From Eastern Finland." *American Journal of Epidemiology.* 128(2) (1988): 370-80.

Kaplan G.A. "Social Contacts and Ischaemic Heart Disease." *Annals of Clinical Research.* 20(1-2) (1988): 131-36.

Karlsson, Jan. *Antioxidants and Exercise.* Champaign, IL: Human Kinetics, 1997.

King Haitung, et al. "Cancer Mortality Among Chinese in the United States " *Journal of Clinical Investigation* 65 (November 1980).

Kramsch, Dieter M. et al. "Reduction of Coronary Atherosclerosis by Moderate Conditioning Exercise in Monkeys on an Atherogenic Diet." *The New England Journal of Medicine* 305 (December 1981): 1483-89.

Lamb, David R. *Physiology of Exercise: Responses & Adaptations.* New York: Macmillan Publishing Company, 1984.

Leon A.S., Connett J., Jacobs D.R., & Rauramaa R., "Leisure-time Physical Activity Levels and Risk of Coronary Heart Disease and Death: The Multiple Risk Factor Intervention Trial." *Journal of the American Medical Association.* 258 (1987): 2388-95.

Leutholtz, B.C. & Ripoll, I. *Exercise and Disease Management. Boca Raton, FL: CRC Press, 1999.*

Lewis, S. F., Taylor, W. F., Graham, R. M., Pettinger, W.A., Shutte, J. E., and Blomqvist, C.G. "Cardiovascular responses to exercise as functions of absolute and relative work load." *Journal of Applied Physiology.* 54(1983):1314-1323.

Lyon, J. L., et al. "Cancer Incidence in Mormons and Non-Mormons in Utah During 1967-1975." *Journal of Clinical Investigation* 65 (November 1980): 1055-61.

MacDougall, J. D., Tuxen, D., Sale, D. G., Moroz, J. R., and Stutton, J. R. "Arterial blood pressure response to heavy resistance training." *Journal of Applied Physiology.* 58(1985):785-790.

Mackinnon, L.T. *Advances in Exercise Immunology.* Champaign, IL: Human Kinetics, 1999.

Markiewics, Konstanty, et al. "Exercise Effect on Certain Biochemical Factors in the Plasma in Untrained Subjects and in Systematically Trained Cyclists." *Acta Physiological Polonica* 31(1980): 325-31.

Masoro E.J. "Biology of aging." *Archives of Internal Medicine.* 147 (1987):166-90.

Maud, P.J. & Foster, C. *Physiological Assessment of Human Fitness.* Champaign, IL: Human Kinetic, 1995.

McArdle, W.D., Katch, F.I., & Katch, V.L. *Exercise Physiology: Energy, Nutrition, and Human Performance.* (4th Ed.) Philadelphia, PA: Lea and Febiger, 1996.

McArdle, W.D., Katch, F.I., & Katch, V.L. *Sports & Exercise Nutrition.* Philadelphia, PA: Lippincott Williams & Wilkins, 1999.

Miller, D. K., and T. E. Allen. *Fitness, A Lifetime Commitment.* Edina, MN: Burgess Publishing, 1986.

Mitchell, J.H. "How to Recognize 'Athlete's Heart'." *The Physician and Sportsmedicine.* 2 20 (August 1992). 87-96.

Morgan, W.P., et al. *Exercise and Mental Health.* Washington, D. C.: Hemisphere Publishing Corp. 1987.

Multiple Risk Factor Intervention Trial Research Group. "Mortality Rates After 10.5 Years for Participants in the Multiple Risk Factor Intervention Trial." *Journal of the American Medical Association.* 263 (1990): 1795-1801.

National Institutes of Health. *Physical Activity and Cardiovascular Health: Consensus Development Conference State.* Washington, D.C., 1995.

Nerem R.M., Levesque M.J., Cornhill J.F. "Social Environment as a Factor in Diet-induced Atherosclerosis." *Science.* 208(4451) (1980): 1475-76.

Ornish, Dean. *Reversing Heart Disease.* New York: Random House. 1990.

Ornish D.M., Scherwitz L.W., Brown S.E., et al. "Adherence to Lifestyle Changes and Reversal of Coronary Atherosclerosis." *Circulation.* 80 (4), (1989): 51-57

Ornish D.M., Schwerwitz L.W.,Brown S.E., et al. "Can Lifestyle Changes Reverse Atherosclerosis?" *Circulation.* 78 (4), (1988): 5-11.

Orth-Gomer K, Unden A.L., Edwards M.E. "Social Isolation and Mortality in Ischemic Heart Disease. A 10-year Follow-up Study of 150 Middle-aged Men." *Acta Medica Scandinavica.* 224(3) (1988): 205-15.

Paffenbarger R.S., Hyde R.T., & Wing A.L. "Physical Activity, All-Cause ortality, and Longevity of College Alumni." *New England Journal of Medicine.* 314 (1986): 605-13.

Patel C., Marmot M.G., Carruthers M., et al. "Trial of Relaxation in Reducing Coronary Risk: Four Follow-up." *British Medical Journal.* 290 (1985): 1103-6.

Pearn, John. "How Long Does It Take to Become Fit." *British Medical Journal* 281 (December 6,1980): 1522-24.

Pena, M., et al. "The Influence of Physical Exercise upon the Body Composition of Obese Children." *Acta Pediatrica Academiae Seicntiarum Hungaricae 9* (1980): 13-21.

Phillips, Roland, et al. "Mortality Among California Seventh-Day Adventists for Selected Cancer Sites." *Journal of Clinical Investigation* 65 (November 1980): 1097-1107.

Pritikin, Nathan. "Optimal Dietary Recommendations: A Public Health Responsibility." *Preventive Medicine* 11 (1982): 733-39.

Public Health Service. *Healthy People 2000: National Health Promotion and Disease Prevention Objectives.* Washington, D.C.: Department of Health and Human Services. PHS 91-50212, 1990.

Rent, Saul. "Diet and Increased Life Span." *Geriatrics* (July 1980): 101-3.

Richard, Denis et al. "Role of Exercise-Training in the Prevention of Hyperinsulinemia Caused by High Energy Diet." *Journal of Nutrition* 112 (1982): 1756-62.

Roberts, S.O., Robergs, R.A., & Hanson, P. *Clincal Exercise Testing and Prescription: Theory and Application.* Boca Raton, FL: CRC Press, 1997.

Sacks F.M., Kass E.H. "Low Blood Pressure in Vegetarians: Effects of Specific Foods and Nutrients." *American Journal of Clinical Nutrition.* 48 (3 Suppl) (1988): 795-800.

Saunders, H. Duane. *Evaluation, Treatment and Prevention of Musculoskeletal Disorders.* Minneapolis: Viking Press, 1985.

Shephard, Roy J. "Nutritional Benefits of Exercise." *The Journal of Sports Medicine and Physical Fitness.* 29(March 1989):83-90.

Siscovick, David S. "Physical Activity and Primary Cardiac Arrest." *Journal of American Medical Association* 248 (1982): 3113-21.

Smith, E. L., and R. C. Serfass. *Exercise and Aging: The Scientific Basis.* Hillside, NJ: Enslow Publishers, 1981.

Strokes, Bruce. "Taking Care of Ourselves." *Environment* 23 (May 1981): 42

Swain J.F., Rouse I.L., Curley C.B., Sacks F.M. "Comparison of the Effects of Oat Bran and Low Fiber Wheat on Serum Lipoprotein Levels and Blood Pressure." *New England Journal of Medicine.* 322 (1990): 147-52.

Tenebaum, Gershon, et. al. "Physical Activity and Psychological Benefits" *The Physician and Sportsmedicine* 20(10) (1992). 180-183.

U.S. Department of Health and Human Services. *Physical Activity and Health: A Report of the Surgeon General.* Sudbury, MA: Jones and Bartlett Publishers, 1998.

Vodak, Paul A. "H D L Cholesterol and Other Plasma Lipid and Lipoprotein Concentrations in Middle-Aged Male and Female Tennis Players." *Metabolism* 29 (August 1980): 745-52.

Walford R.L., Harris S.B., Weindruch R. "Dietary restriction and aging." *Journal of Nutrition.* 117 (1987): 1650-54.

Warren, Michelle P. "The Effects of Exercise on Pubertal Progression and Reproductive Function in Girls." *Journal of Clinical Endocrinology and Metabolism* 51 (1980): 1150.

West, D. W., et al. "Cancer Risk Factors: An Analysis of Utah Mormons and Non-Mormons." *Journal of Clinical Investigation* 65 (November 1980): 1083-95.Westcott, Wayne L. *Strength fitness: Physiological Principles and Training Techniques.* Boston: Allyn and Bacon, 1982.

Williams R.B. "The Role of the Brain in Physical Disease: Folklore, Normal Science, or Paradigm Shift?" *Journal of American Medical Association.* 263 (1990): 1971-2.

Wilmore, J.H. & Costill, D.L. *Physiology of Sport and Exercise.* (2nd Ed.) Champaign, IL: Human Kinetics, 1999.

Winter, W. W. "Effect of Endurance Training on Liver and Response to Prolonged Submaximal Exercise " *American Journal of Physiology* 240 (May 1981): 330-34

Index